# Springer Complexity

Springer Complexity is an interdisciplinary program publishing the best research and academic-level teaching on both fundamental and applied aspects of complex systems – cutting across all traditional disciplines of the natural and life sciences, engineering, economics, medicine, neuroscience, social and computer science.

Complex Systems are systems that comprise many interacting parts with the ability to generate a new quality of macroscopic collective behavior the manifestations of which are the spontaneous formation of distinctive temporal, spatial or functional structures. Models of such systems can be successfully mapped onto quite diverse "real-life" situations like the climate, the coherent emission of light from lasers, chemical reaction-diffusion systems, biological cellular networks, the dynamics of stock markets and of the internet, earthquake statistics and prediction, freeway traffic, the human brain, or the formation of opinions in social systems, to name just some of the popular applications.

Although their scope and methodologies overlap somewhat, one can distinguish the following main concepts and tools: self-organization, nonlinear dynamics, synergetics, turbulence, dynamical systems, catastrophes, instabilities, stochastic processes, chaos, graphs and networks, cellular automata, adaptive systems, genetic algorithms and computational intelligence.

The three major book publication platforms of the Springer Complexity program are the monograph series "Understanding Complex Systems" focusing on the various applications of complexity, the "Springer Series in Synergetics", which is devoted to the quantitative theoretical and methodological foundations, and the "SpringerBriefs in Complexity" which are concise and topical working reports, case-studies, surveys, essays and lecture notes of relevance to the field. In addition to the books in these two core series, the program also incorporates individual titles ranging from textbooks to major reference works.

# Springer Series in Synergetics

**Founding Editor: H. Haken**

The Springer Series in Synergetics was founded by Herman Haken in 1977. Since then, the series has evolved into a substantial reference library for the quantitative, theoretical and methodological foundations of the science of complex systems.

Through many enduring classic texts, such as Haken's *Synergetics and Information and Self-Organization*, Gardiner's *Handbook of Stochastic Methods*, Risken's *The Fokker Planck-Equation* or Haake's *Quantum Signatures of Chaos*, the series has made, and continues to make, important contributions to shaping the foundations of the field.

The series publishes monographs and graduate-level textbooks of broad and general interest, with a pronounced emphasis on the physico-mathematical approach.

More information about this series at http://www.springer.com/series/712

Juan C. Vallejo • Miguel A.F. Sanjuan

# Predictability of Chaotic Dynamics

## A Finite-time Lyapunov Exponents Approach

 Springer

Juan C. Vallejo
Department of Physics
Universidad Rey Juan Carlos
Móstoles, Madrid, Spain

Miguel A.F. Sanjuan
Department of Physics
Universidad Rey Juan Carlos
Móstoles, Madrid, Spain

ISSN 0172-7389                    ISSN 2198-333X    (electronic)
Springer Series in Synergetics
ISBN 978-3-319-84771-9            ISBN 978-3-319-51893-0    (eBook)
DOI 10.1007/978-3-319-51893-0

Printed on acid-free paper

This Springer imprint is published by Springer Nature
The registered company is Springer International Publishing AG
The registered company address is: Gewerbestrasse 11, 6330 Cham, Switzerland

The original version of this book was revised. An erratum to this book can be found at DOI 10.1007/978-3-319-51893-0_5

*To my wife Laura and my children Alicia and Pedro*

*To my wife Céline and my daughters Alicia and Mónica*

# Preface

Mankind has always been concerned with the desire of understanding the universe, knowing the ultimate reasons behind past events and having the ability of forecasting the future ones. In a very simplified view, the main task of a physicist is to observe the nature, to build models and to derive predictions from them. But, in some fields, as, for instance, astronomy, one recollects the necessary information from observed objects without the possibility of having direct access to them, so one has not the possibility of altering the key parameters of the studied objects. Even worst, the timescales applicable may be out of the human timescales. A key issue in these cases is to study the subject of observation through numerical simulations.

In recent years, simulations have gained much relevance in physics. They play an important role as a new tool in addition to theory and experiments. Furthermore, they can be very useful in exploring the consequences of varying the parameters in physical models. With the widespread usage of computer simulations to solve complex dynamical systems, the reliability of the numerical calculations is of increasing interest. We can take a model, or set of equations describing the system, and integrate it during a certain time interval. The question to answer here is: how valid is the resulting forecast? Every model has inherent inaccuracies leading its results to deviate from the true solution. Any numerical schema used for solving it will introduce several errors. Round-off errors are present because it is impossible to represent all real numbers exactly on a machine with finite memory. Truncation errors are committed when the iterative method is terminated or a mathematical procedure is approximated and the approximate solution differs from the exact solution. Or discretisation errors must be taken into account when the solution of the discrete problem does not coincide with the solution of the continuous problem.

The reliability of the calculations is directly related to the regularity and instability properties of the modelled flow. This is an interdisciplinary scenario, where the underlying physics provides the simulated models, nonlinear dynamics provides their chaoticity and instability properties and the computer sciences provide the actual numerical implementation.

This book faces the problem of characterising the time (predictability time) when a numerical prediction can be considered valid by using techniques and concepts derived from the nonlinear dynamics and chaos theory.

Because of the possible differences between the real problem and the model used for making predictions, and because the numerical methods will introduce different errors and perturbations, the resulting numerical solution of a model will not match with the real one.

A system is said to be chaotic when it presents strong sensitivity to initial conditions. It is obvious that the presence of chaos can impose certain limits to the time when two initially closed trajectories, the real one and the calculated one, will remain close. Chaos does not always imply a low predictability. An orbit can be chaotic and still be predictable, in the sense that the chaotic orbit is followed, or shadowed, by a real orbit, thus making its predictions physically valid. The computed orbit may lead to right predictions despite being chaotic because of the existence of a nearby exact solution. This true orbit is called a shadow, and the existence of shadow orbits is a very strong property.

There are several books that deal with the selection of the most suitable numerical scheme for solving a given problem, others that describe different chaos indicators for characterising the presence of chaos and others that perform a thoroughly theoretical study of the underlying shadowing theories.

This book aims to take a different approach and it performs a descriptive analysis of how one can gain insight in the study of the predictability of a system. This characterisation will be done through the computation of the finite-time Lyapunov exponents and their distributions. This book will circumscribe to the field of the dynamical continuum flows, even when maps and discrete systems are mentioned when needed as part of the discussion.

As a consequence, this book presents basic concepts on dynamical flows needed for the computation of asymptotic and finite-time Lyapunov exponents. It will show how this computation can provide different properties of a given system related to its predictability.

This book approaches these issues from a numerical exploration perspective. The mathematical background is not detailed, but just introduced and appropriate bibliographic references are given to the interested reader. It does not focus on the mathematical derivations of required well-known algorithms and schemes. Conversely, it describes how to use them for the purpose of this work, that is, analysing the predictability of a given system. So, it focus, on describing the procedures required for carrying out these analyses in a step-by-step method. After discussing these procedures, we present in each chapter a reduced set of simple case studies on conservative and dissipative systems.

This book is primarily developed as a text at the postgraduate level and also as a reference book for researchers working and/or interested in the field of the predictability of dynamical systems. It is a self-contained book, where all needed techniques are properly described so that a reader is able to reproduce the results presented and may apply them to any problem of his/her interest.

This work is structured into four main chapters and one appendix. This monograph has been written in a way that every chapter can be read independently of the others. This self-consistency means that some overlap may be found when reading the whole book, but this overlapping will allow the interested reader to directly consult some sections and methods described in a given section of the book.

The book begins with an introduction to the ideas of forecasting in science in order to give a historical perspective. This introductory part discusses the forecasting and its relationship with the predictability of the numerical computations. The second chapter is devoted to the calculus of the finite-time Lyapunov exponents and describes how their distributions provide information of the properties of a dynamical flow at local scales. The third chapter expands the previous ideas, and presents how the distributions of finite-time Lyapunov exponents change as the different regimes of the dynamical flow are reached, and how these changes can be used for characterising the system. The final chapter is concerned with the calculus of the predictability itself, in terms of presence of the shadowing property, as indicator of the reliability of any results obtained from solving numerically a given dynamical system. The appendix describes the main algorithms used along the book for computing the finite-time Lyapunov exponents, with some comments about how to implement and use them in the most efficient way.

We are indebted to Ricardo L. Viana, Juergen Kurths and James A. Yorke for their fruitful comments and discussions concerning the computation of finite-time Lyapunov exponents at very small local scales. Finally, we acknowledge the support and encouragement given by our family members during the preparation of this monograph.

Móstoles, Madrid, Spain                                                 Juan C. Vallejo
Móstoles, Madrid, Spain                                        Miguel A.F. Sanjuan
November 2016

# Contents

**1 Forecasting and Chaos** .......................................................... 1
  1.1   Historical Introduction .................................................. 1
      1.1.1   The Scientific Method ........................................ 1
      1.1.2   Forecasting and Determinism .............................. 3
  1.2   Chaotic Dynamics ...................................................... 9
  1.3   Computer Numerical Explorations .................................... 13
      1.3.1   Solving ODEs Numerically.................................. 13
      1.3.2   Numerical Forecast .......................................... 14
      1.3.3   Symplectic Integrators ...................................... 17
  1.4   Shadowing and Predictability .......................................... 19
  1.5   Concluding Remarks ................................................... 22
  References ...................................................................... 22

**2 Lyapunov Exponents** ........................................................... 25
  2.1   Lyapunov Exponents .................................................... 25
  2.2   The Lyapunov Spectrum ............................................... 27
  2.3   The Lyapunov Exponents Family....................................... 30
  2.4   Finite-Time Exponents ................................................. 33
  2.5   Distributions of Finite-Time Exponents ............................... 34
  2.6   The Harmonic Oscillator ............................................... 36
  2.7   The Rössler System .................................................... 38
  2.8   The Hénon–Heiles System ............................................. 43
  2.9   Concluding Remarks ................................................... 54
  References ...................................................................... 56

**3 Dynamical Regimes and Time Scales** ......................................... 61
  3.1   Temporal Evolution .................................................... 61
  3.2   Regimes Identification ................................................. 63
  3.3   Transient Behaviours, Sticky Orbits and Transient Chaos ........... 64
  3.4   The Hénon-Heiles System ............................................. 65
  3.5   The Contopoulos System ............................................... 71
  3.6   The Rössler System .................................................... 80

3.7   Hyperbolicity Characterisation Through Finite-Time
      Exponents ................................................................. 82
3.8   Concluding Remarks ..................................................... 86
References ...................................................................... 87

4  **Predictability** ................................................................. 91
   4.1   Numerical Predictability ............................................... 91
   4.2   The Predictability Index ............................................... 93
         4.2.1   The Hénon-Heiles System ..................................... 96
         4.2.2   The Contopoulos System ..................................... 105
         4.2.3   The Rössler System .......................................... 106
         4.2.4   A Galactic System ........................................... 112
   4.3   Concluding Remarks .................................................. 123
   References ................................................................... 126

**Erratum to: Predictability of Chaotic Dynamics: A Finite-time
Lyapunov Exponents Approach** ............................................. E1

A  **Numerical Calculation of Lyapunov Exponents** ......................... 129
   A.1   The Variational Equation .............................................. 129
   A.2   Selection of Initial Perturbations ..................................... 132
   A.3   Other Methods ........................................................ 134
   A.4   Practical Implementation for Building the Finite-Time
         Distributions .......................................................... 135
   References ................................................................... 136

# Acronyms

We have tried to reduce the number of abbreviations used along the text. However, there is a variety of notations and acronyms found in the related literature and here we summarise them, as well as the most frequently used symbols found along the text.

AFTLE  Averaged Lyapunov exponent
DLE   Direct Lyapunov exponent
ELN   Effective Lyapunov number
FTLE   Finite-Time Lyapunov exponent
FSLE   Finite-Size Lyapunov exponent
LN    Lyapunov number
LCE   Lyapunov characteristic exponent
LCI    Lyapunov characteristic indicator
LCN   Lyapunov Characteristic number
MLE   Maximal Lyapunov exponent
ODE   Ordinary Differential equation
SDLE   Scale-dependant Lyapunov exponent
SVD   Singular value decomposition
UDV   Unstable dimension variability
UPO   Unstable periodic orbit

$\chi$    Finite-time or short-time Lyapunov exponent, FTLE
$\Delta t$   Finite-time interval length
$\phi(\boldsymbol{x}, t)$  Solution of the flow equation
$P_+$   Probability of positivity
$h$    Predictability index
$J$    Jacobian matrix
$\lambda$    Asymptotic Lyapunov exponent

# Chapter 1
# Forecasting and Chaos

## 1.1 Historical Introduction

### 1.1.1 The Scientific Method

Mankind has always been concerned with the desire of understanding the Universe, knowing the ultimate reasons behind past events, and having the ability of forecasting the future ones. From the earliest times, the study of natural cycles has been needed for a successful harvest. The Astronomy, as one of the oldest sciences, was born with the main task of compiling the several observed phenomena in the skies. It attempted to understand the underlying mechanisms of the observations and trying to figure out what was going to be observed in the future.

In a very simplified view, the main task of a physicist today is still to observe the nature, to build models from those observations, and to use them for deriving predictions. Forecasting is the process of making predictions of the future based on past and present data. Thanks to the scientific models we are able to understand the nature and how it works, forecast the weather, anticipate the position of celestial bodies, send probes to them, understand the behaviour of molecules and atoms, build new materials, imagine new ways for generating energy, etc.

Nevertheless, the importance of forecasting goes beyond the practical purposes and points to the essence of the scientific method itself. A given physical model must capture the reality as faithfully as possible and constitutes our understanding of the ultimate reasons for the physical processes under study. Precisely, one of the key aspects of the scientific method is the possibility of making predictions using the selected model and to confront these predictions with new observations.

---

The original version of this chapter was revised.
An erratum to this chapter can be found at DOI 10.1007/978-3-319-51893-0_5

© Springer International Publishing AG 2017
J.C. Vallejo, M.A.F. Sanjuan, *Predictability of Chaotic Dynamics*,
Springer Series in Synergetics, DOI 10.1007/978-3-319-51893-0_1

This confrontation constitutes a test of the goodness of the model, that is, of our understanding of reality.

As a consequence, a key issue is the falsifiability or refutability of the theory, or inherent possibility that it can be proved to be true or false. The model is built upon on existing observations but is used for making predictions which can be confronted with the reality. One must analyse whether it is possible to conceive of an experiment leading to new observations which may negate that model. And it is out of question that such a new experiment must be done. If a physical theory is not falsifiable, it cannot be considered valid, because it will not allow to prove its validity.

The validity of a model, based on certain assumptions is only of temporal nature. The fact that a theory may have successfully survived several times the refutability test does not circumvent that it may eventually fail. Past and present successes do not imply future successes. Noticeably, these failures must be seen as positive facts, because they lead to a better definition of the boundaries in the application of the physical law, and may help to refine its underlying assumptions. In some cases, the failure may lead to a complete replacement of the old model by a new set of laws, which in turn may open news ways of knowledge.

A famous example of the above is that of Newton's law of universal gravity. When formulated in 1687, these laws allowed to successfully explain the motion of the planets already known from ancient times, from Mercury up to Saturn. Then, the planet Uranus was discovered by the British astronomer Frederick William Herschel in 1781. This was the very first planet discovered in modern times, and it was discovered by chance, because no one expected to find a new planet at that time. It had been observed several times in the past, but Herschel was the first in properly annotating and observing it, realising about the true nature of Uranus as a moving planet and not as a fixed star.

Once it was realised that it was a genuine planet, not a star neither a comet, subsequent observations were performed along the following decades. They revealed substantial deviations from the tables based on predictions done by the formulation of the Newton's law of universal gravity. So confident was the scientific community in the goodness of the Newton's laws, that it was hypothesised that an unknown body was perturbing the orbit through gravitational interaction. The position of this body was predicted in 1846 by the French mathematician Urbain Jean Joseph Le Verrier.[1] Some observations were done in the area predicted by Le Verrier and the planet was finally found. The discovery of Pluto roughly followed a similar path. Careful analyses of the orbit of the recently discovered planet Neptune, and also, again, of Uranus, predicted the existence of another new planet. These discrepancies would ultimately lead to the discovery of Pluto by the American astronomer Clyde Tombaugh in 1930. All these successes gave strong confidence in the infallibility of the Newton's law of universal gravitation.

---

[1] In 1843, the British mathematician and astronomer John Couch Adams also began to work on the orbit of Uranus using the data he had, and he has been sometimes credited by the discovery of Neptune.

Some years later, in 1859, Le Verrier also published a detailed study of Mercury's orbit and how its perihelion advanced by a small amount each orbit. But, again, there were discrepancies between the observed and the predicted data. Following the same, previously successful, reasoning, Le Verrier hypothesised that these discrepancies were the result of the presence of a small planet inside the orbit of Mercury. There was such a confidence in the Newton's classical laws, that the new planet, not yet discovered was even given a name, Vulcan.

Several claims for the intra-mercurial planet followed, but none of them were confirmed. Conversely, the arrival of the Einstein's theory of relativity in 1915 predicted exactly the observed amount of advance of Mercury's perihelion. The new theory indeed modified the predicted orbits of all planets when compared with the Newtonian theory. But the magnitude of these differences was only most evident for the closest planet to the Sun, Mercury.

Does this mean that the Newton's laws are not valid any longer? Does this mean that the Einstein's laws are the only real truth for explaining the Universe? It may appear a new theory, not yet discovered that may replace Einstein's theory? Both Newton's and Einstein's laws can be considered valid, but only when one takes into account their respective applicability constraints.

## 1.1.2   Forecasting and Determinism

Forecasting is then one of the pillars of the scientific method. Risk and uncertainty are central to forecasting and prediction. As a consequence, the estimation of the degree of uncertainty attached to a given forecast is a key issue. While qualitative forecasting techniques are somehow subjective, quantitative forecasting models may predict future data as a function of past data. Dynamical models can be seen as a way for making quantitative forecasting.

A dynamical model is the generic name typically described mathematically by differential equations, which are obtained by the analysis of the physical system at a fundamental level, yet involving approximations sufficient to simplify the model [58]. They constitute a mathematical rule for the time evolution on a state space, characterised by the coordinates of the phase space that describe the state of the system at any instant.

This dynamical rule specifies the immediate future of all state variables, given only the present values of the same state variables. Up to the end of nineteenth century, science relied on the concept of determinism. Such a concept was nicely described by the French mathematician and astronomer Pierre Simon Laplace in 1776, in his book "A philosophical essay on probabilities" [45],

> We ought then to regard the present state of the universe as the effect of its anterior state and as cause of the one which is to follow. Given for one instant an intelligence which could comprehend all the forces by which nature is animated and the respective situation of the beings who compose it -an intelligence sufficiently vast to submit these data to analysis- it would embrace in the same formula the movements of the greatest bodies of the universe and those of the lightest atom; for it, nothing would be uncertain and the future, as the past, would be present to its eyes.

And it follows,

> The human mind offers, in the perfection which it has been able to give to astronomy, a feeble idea of this intelligence. Its discoveries in mechanics and geometry, added to that of universal gravity, have enabled it to comprehend in the same analytical expressions the past and future states of the system of the world.

At the time those lines were written, it was already known that an absolute error free precision in the initial measurements was unreachable in practice. But it was thought that close enough initial conditions would lead to final similar solutions, close enough as well. It seems rather natural to think that with an adequate increase in numerical computational facilities, the errors could be neglected and that from a set of initial conditions known with enough precision, one could predict the future state of a dynamical system.

With the advent of the study of nonlinear systems, the previous solid scheme begun to be modified. When nonlinearity is present, deterministic dynamics can lead to very complex dynamics. And one prediction can be destroyed by an initial error, even by a very small one. The first work where this view was exposed was a memory of the French mathematician and physicist Henri Poincaré, "On the three-body problem and the equations of dynamics" [36]. This work was a consequence of the prize of 2500 kronas that the King Oscar II, King of Sweden and Norway, offered in 1887 to whom may provide an answer to the practical question "Is the Solar System stable?". Poincaré's efforts in solving this question not only lead him to win the prize, but lead to the birth of one of the most fruitful branches of the mathematics, the Topology, that he called Analysis Situs. These initial ideas were later expanded in his book "Les Méthodes nouvelles de la mécanique céleste", published in three parts between 1892 and 1899 [37].

Poincaré first established the properties of the necessary dynamical equations. Then, he applied the results to the problem of an arbitrary number of bodies under Newtonian gravitational interaction. Finally, he analysed the existence of periodic solutions, following the classic approach of developing the necessary variables as infinite series, and finding that there are series with periodic coefficients formally satisfying the equations of motion. But he did not attempt to prove the convergence of the series. Conversely, he proved the existence of periodic solutions using another different approach. The existence of these solutions proved, in turn, the convergence of the series.

His work introduces by the first time the nowadays commonly used Poincaré section technique. He defined here for the first time what a consequent is (see Fig. 1.1), and defined an invariant curve as any curve that is its own consequent. He also introduced how the Poincaré section is associated with a map, the Poincaré map. Poincaré used these tools for presenting a detailed discussion about the possible existence of solutions of different types. When discussing the doubly asymptotic solutions in the third volume of his book [37], he constrained his analysis to the special case of zero mass of the perturbed planet, circular orbit of the perturbing planet and zero inclinations, that is, the reduced Hill's problem. And in the section

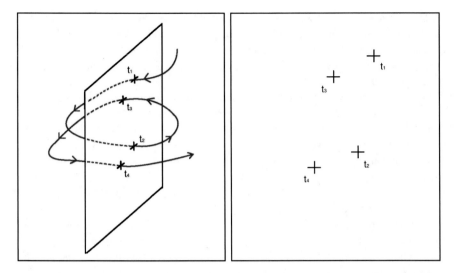

**Fig. 1.1** The Poincaré section is built by selecting an arbitrary plane in the phase space of the problem under study. The technique relies on analysing how the trajectory intersects such a plane. Every intersection point in the plane, or surface of section is a consequent of the trajectory. The time between consecutive crosses is variable, typically labeled as $T_{\text{cross}}$. One must take into account the sign of the velocity when crossing the plane from one subspace to the other, and vice versa. One can also compute an averaged $< T_{\text{cross}} >$ for all consequents. For a periodic orbit, it will be equal to the period of the movement

devoted to the existence of homoclinic solutions in this problem, he wrote:

> 397. When we try to represent the figure formed by these two curves and their intersections in a finite number, each of which corresponds to a doubly asymptotic solution, these intersections form a type of trellis, tissue, or grid with infinitely serrated mesh. Neither of the two curves must ever cut across itself again, but it must bend back upon itself in a very complex manner in order to cut across all the meshes in the grid an infinite number of times.

And it follows,

> The complexity of this figure will be striking, and I shall not even try to draw it. Nothing is more suitable for providing us with an idea of the complex nature of the three-body problem, and of all the problems of dynamics in general, where there is no uniform integral and where the Bohlin series are divergent.

The discovery of Poincaré means that one can find extremely complex behaviours even in systems as simplified as the reduced Hill's problem. One trajectory starts in one point of intersection of the cross section, and the following time, will appear in other different point, a different consequent of the grid. And so on and so forth. But the grid is so stretched and folded that the points seem to follow a complex random pattern.

The physical systems which show this behaviour of strong sensitivity to initial conditions, making them unpredictable, are nowadays called "chaotic systems".The American meteorologist Edward Lorenz published his article "Deterministic Non-periodic flow" [29] in 1963, where he presented many of the ideas that today belong to the chaos theory.

He started computing a simplified version of the equations describing the Bénard convection. But even being a very simplified version, they could not be solved analytically, and a numerical approach was needed. Lorenz found that the behaviour of the trajectories was very complex, without periodic motions. He found orbits showing violent oscillations around two points, resembling a random, unpredictable motion. The object traced by these trajectories is a "strange attractor", as it was labeled later on by the Belgian-French mathematical physicist David Ruelle and the Dutch mathematician Floris Takens in 1971 [39].

Lorenz also discovered another interesting fact. He used a computer, a Royal McBee LGP-300, for doing his analyses. Because his computer was very slow, and because he wanted to study the trajectory during a very long time interval, he started a second simulation in the middle point of a previous one. As initial point of the second simulation he took the middle point coordinates as directly read from a printer output.

He observed that the second integration behaved very similar to the first one for a while. But later on, its behaviour was completely different. He realised that the reason was that the printed numbers were a rounded version of the memory-allocated point used in the first simulation. He had done that presuming there should not be a major difference in the prediction using an initial condition or another one very close to it. He had found, and described what is popularly known as the "butterfly effect", or strong sensitivity to initial conditions that characterises any chaotic system.

One year before the publication of Lorenz's paper, 1962, the French astronomer Michel Hénon studied the motion of stars in a simplified Galaxy model, with a behaviour dependent on the energy level. Up to that date, classic tools used for doing so, as the theory of perturbations, relied on the fact that all solutions were supposed to be periodic or separable in several periodic or quasi-periodic components. He started to study his model with a recent graduate student, the American Carl Heiles. Again, they used a computer for doing the calculation, which was not a common tool at that time.

Their results were published in the article "The applicability of the third integral of motion: some numerical experiments" published in 1964 [18]. They showed how the system behaves at low energy levels, with the trajectories regular and periodic. In this case, the solutions of the original model and the solutions based on a series expansion, which is equivalent to approximate the non-integrable problem by an integrable system, were very similar. But at energy levels above certain threshold, the real solutions showed a Poincaré section based on erratic, random-like points, fulfilling all the available phase space. We can see the clear differences between the

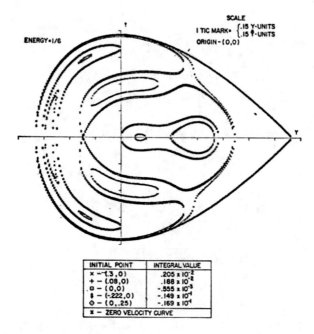

**Fig. 1.2** Poincaré section for the Hénon-Heiles system and Energy =1/6 showing solutions based on series expansion of the original model. Taken from [12] © AAS. Reproduced with permission

orbits in the real system and in the approximated system, that continues providing regular solutions by comparing Figs. 1.2 and 1.3, directly taken from [12].

In 1975, the Chinese-American mathematician Tien-Yien Li and the American mathematician James A. Yorke published the classic paper "Period Three Implies Chaos" [28], where the term "chaos" was used for the first time in the dynamical systems literature. Finally, the concept of "chaos control" was born in 1990, as a branch of chaos theory pioneered by James A. Yorke, with the American physicist Edward Ott and the Brazilian physicist Celso Grebogi. Today, the chaos theory is a mature branch of physics that is successfully applied to a wide set of fields.

We have constrained the discussion so far to the field of classical dynamics. And we can extend this discussion by entering in the realm of quantum mechanics. The presence of chaos may jeopardise the determinism in a practical way, but such a presence does not forbid the existence of a deterministic Universe by itself. This means that the future of the Universe may be already predetermined even when we would be unable to reach any knowledge about it. The future could be already written since the Big Bang, even when we will be unable to compute it. The so-called strong determinism goes even further, and claims that it is not only the history of all our Universe what is written, but the history of all possible Universes that must be present when taken into account Quantum Mechanics.

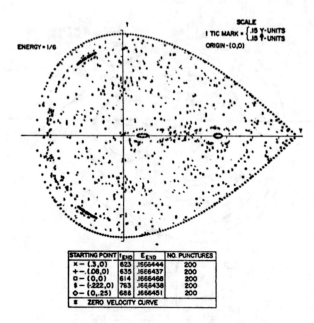

**Fig. 1.3** Poincaré section for the Hénon-Heiles system and Energy = 1/6, solutions of the original model. Taken from [12] © AAS. Reproduced with permission

We have seen that classical dynamics keeps the determinism of the Universe, even when the presence of chaos may also break its computability from a practical point of view. Quantum Dynamics may introduce additional factors to take into account. Following the ideas of the British physicist Roger Penrose [34], Quantum Dynamics can be considered formed by two parts. One is purely deterministic, formed by the fully deterministic Schrödinger equation about the evolution of the wave function $\Psi$. But there is another process, responsible for the provision of the state vector reduction and collapse of the wave function. This latter process takes the squares of the quantum amplitudes for obtaining classical probabilities, and it is this second process which is responsible for adding uncertainties and probabilities in Quantum Theory. This is a probability process that is not continuous and cannot be derived from the fully deterministic process. Whether it is indeed a random process or not is still under debate. However, if that is the case, or even if there is any new unknown process linking both, it could be essentially non-algorithmic. That would mean the future would completely loss its deterministic nature and its future states will never be computed from the present ones.

## 1.2 Chaotic Dynamics

Dynamical systems forecast future states based on a set of initial conditions. Dynamical systems are deterministic if there is a unique well-identified consequent to every state. Conversely, they are labeled as random or stochastic if there is a probability distribution of possible consequents.

Our work will focus on the study of deterministic dynamical systems with a limited number of degrees-of-freedom, sometimes very small. In systems with many degrees-of-freedom, such as those found when studying the movements of molecules in a gas, it is impossible to manage the huge amount of data required for a detailed description of a single state. One must use laws for averaged quantities, and to assign different probabilities to each variable and to the state of every individual particle. Because of that, these dynamical systems with many degrees-of-freedom are sometimes labeled as stochastic systems, even when they are not intrinsically random processes, but subject to fully deterministic equations. They are analysed with statistical tools just because the analyses done using individual deterministic equations are too far from the practical computing capabilities.

The term predictability refers to the assessment of the likely errors in a forecast, either qualitatively or quantitatively. Indeed, random processes can be considered predictable processes if it is possible to know the next state at the present time. One may think that deterministic systems are always predictable. But, as we have seen in the historical introduction, the study of nonlinear systems leads to the discovery that even in deterministic systems, the predictability can deteriorate, and very fast, with time. These systems are called chaotic systems.

The chaotic systems are linked to the notion of complex systems. Conversely to common systems, where we can obtain a realistic description of the system by dissecting it into its structural component parts, complex systems require a different approach due to the nonlinear interactions. At best, the result of the individual, per-component approach does not substantially add anything to our understanding or at worst it can even be misleading. A global view or holistic approach is then needed in these systems.

A chaotic system is not a stochastic system. It is a deterministic system that shows a strong sensitivity to the initial conditions. This implies that they can be low-dimensional with a very low number of degrees-of-freedom. Independently of that, they show a complex, random-like behaviour. So, the identification of chaotic orbits is a key element in the analysis of complex systems.

We should note that the presence of chaos is a concept different from the fact that an orbit is stable or not. The stability property characterises whether nearby (i.e. perturbed) orbits will remain in a neighbourhood of that orbit or will move away from it. There are two types of stability, a weaker and a stronger one. The first type is marginal stability. This is when every orbit starting in the neighbourhood of a specified orbit will remain in its neighbourhood at the same distance. The other type is the asymptotic stability. This happens when every orbit starting in the neighbourhood of a specified orbit will approach the specified orbit asymptotically. Conversely, an orbit is said to be unstable when every orbit starting in the neighbourhood of a specified orbit will leave exponentially its neighbourhood. There is a variety of criteria used for characterising the stability of an orbit, and the interested reader is pointed, for instance, to [1].

Observing a complex trajectory in the phase space of a deterministic system can be a sign of strong sensitivity to initial conditions, but it is not a synonym of chaotic motion. There are certain systems where the behaviour of a system cannot be predicted, but there is a final state where two or more exclusive states coexist (multistability), or there is no final state defined, leading to functions that do not repeat values after some period, showing aperiodic or irregular motions, but without showing exponential divergence of initial nearby trajectories.

Despite the above, the presence of complex motions is the obvious first step for detecting the presence of chaos. Unfortunately, plotting the orbits of a dynamical system with more than two dimensions is not a straightforward task. In dissipative systems, the trajectories tend to a certain confined region of the phase space, and this region is named "attractor". From the Poincaré-Bendixon theorem, in continuous flows of second order, attractors can only be of four types: sources, sinks, connexion arcs and limit cycles. But in higher dimensional systems, the attractors can be more complex. They are named in these cases "strange attractors", and they usually present a fractal geometry. The presence of strange attractors in a system is a clear sign of the intrinsic complexity of the motion.

There are certain techniques such as the Poincaré section that may reduce the number of dimensions to visualise in flows with a number of dimensions higher than two. The Poincaré section, when chosen carefully, can be used to accurately show the stable and unstable regions of phase space. The inspection of the areas where there are closed curves or, conversely, randomly scattered dots, or areas where the cumulants get crowded as the integration progresses, provides insight into the nature of the system. But the selection of a proper surface of section is still an issue, and the results may strongly depend on the final choice.

There are other alternatives. The search for invariants is one of the most common tools for understanding the dynamics of non-integrable systems (see a classical example in [9]). Usually, the basic building blocks of the dynamics, the fixed points and associated periodic orbits, are located and studied. Later on, their surroundings are analysed, as the stable orbits are mostly surrounded by quasi-periodic orbits, while unstable periodic orbits by chaotic orbits. But the complexity of the higher order orbits makes this procedure less straightforward.

Moreover, the complexity of the algorithms for searching periodic orbits increases with the number of dimensions. Independently of using symplectic schemes for the integration of a given orbit, some typical algorithms such as those based on Newton algorithms must explore a set of initial conditions in the phase space keeping the exploration within the initial energy subspace. The computation of the invariant tori and invariant manifolds also gains a high degree of complexity in realistic models.

One common sign pointing to the presence of chaos is the aperiodicity of the orbit. One periodic orbit will contain one or more frequencies that are rationally related (resonance). One quasi-periodic orbit will contain at least two incommensurable frequencies, that is, the ratio of the frequencies is an irrational number. These orbits are labeled as regular orbits. A regular orbit contained in $N$ dimensions can be decomposed into $N$ independent periodic motions. They are subject to a separable potential, and the orbit can be described as a path on an invariant $N$-dimensional torus. The aperiodic orbits are orbits whose motion cannot be described as the sum of periodic motions as $t \rightarrow \infty$. These irregular orbits cannot be decomposed into independent periodic motions, and they can move anywhere energetically permitted. Although it is generally assumed that irregular orbits and chaotic orbits are the same in Hamiltonian systems, this has not been proven in general [5].

As a consequence of the above, one common approach for chaos detection is the Fourier Analysis. It starts from the fact that any periodic function can be expressed in form of a Fourier series, based on one fundamental frequency and harmonically related frequencies. Quasi-periodic functions show a power spectrum with more than one fundamental frequency, and their linear combinations with integer coefficients. So, a typical signal of a regular movement is a discrete power spectrum. Conversely, aperiodic functions show a continuum spectrum. And, in low-dimensional systems, this is a good indicator of the presence of chaos.

This approach can be extended by performing a spectral analysis of the projections of the orbit on the principal axes. For instance, we have three frequency plots for a three-dimensional system. The spectral analysis method relies on detecting resonances between frequencies on different axes to identify different families of orbits. There are different choices for the spectral analysis algorithm and the choice of which data, as positions or velocities, but the basics are the same in all cases. See [5, 32, 55] for further details.

Another family of methods for detecting the presence of chaos is built upon the Wavelet analysis. This technique aims at representing a signal in the time-frequency domain by performing a frequency analysis of the signal at different levels, or "scales", of detail. Some methods focus on detecting instantaneous events or peaks, and others focus on detecting the evolution of the major frequencies of the system, the instantaneous frequencies. In [4] a method is described to detect these instantaneous frequencies called "ridges", and in [7], this method is applied to detect chaotic behaviour for time series.

Another indicator of chaos is the use of correlation functions, that indicates if a given coordinate conserves a given relationship of the previous value with the current value, once a given time $\tau$ has elapsed. This time interval is called correlation time. The correlation function, sometimes called self-correlation, will indicate how a given value in a time series is similar to its value several intervals later. A correlation function that grows faster with the number of intervals is a clear indication of chaos. Conversely, one function that decreases, or decreases after a growing transient, can indicate a regular movement. Indeed, there is a relationship between a faster growing correlation function and a broad, almost continuum power spectrum.

The correlation function can be also calculated using the concept of invariant density $\rho(x)$, and this concept also appears frequently when searching for signs of the presence of chaos. An invariant measure is defined as a measure that is preserved by some function. The Krylov–Bogolyubov theorem proves the existence of invariant measures under certain conditions on the function and space under consideration. If one initially sets in the phase space a given number of systems with a density proportional to $\rho(x)$, that density will not vary as the flow evolves. In principle, there can be several functions fulfilling that property. When there is only one, the system is called ergodic. This means that every trajectory goes arbitrarily close of any point of the accessible phase space. When the system is not ergodic, one can define different invariant densities, but the presence of ergodicity is a sign of the presence of chaos.

There is a whole set of methods relying on the study of the deviation vectors aiming at detecting the presence of chaos in a system. Those methods rely on considering the vector between the position of the particle to be observed and the position of a slightly displaced particle. This vector is called a deviation vector. Every method aims to assign a quantitative measure for the type of orbit to these deviation vectors. The Lyapunov exponents are the best-known example of this method.

The Lyapunov exponents measure the average rate of divergence (or convergence) of two initially close trajectories. Actually there is not only one exponent, but there are $N$ Lyapunov exponents in a system with an $N$-dimensional phase space. When "the Lyapunov exponent" is mentioned, we typically mean the largest one, pointing to the fastest growing direction.

The asymptotic Lyapunov exponents provide an indication on the globally averaged chaoticity of the system during an infinite integration time. But while they measure the asymptotic divergence of infinitesimally neighbouring trajectories, it is not always possible or desirable to perform these very long integrations and the limit value. Indeed, sometimes the asymptotic limit, thus the exponents themselves, may not exist [31].

The standard definition of the Lyapunov exponent uses a very long (infinite) convergence time. Due to the sometimes slow convergence towards the asymptotic value, many other numerical indexes and fast averaged indicators have been developed. We can cite, among others, the Rotation index [57], the Smaller alignment index [46] or its generalisation, the Generalised alignment index [48], the Mean exponential growth factor of nearby orbits [8] the Fast Lyapunov indicator [10], the Relative Lyapunov indicator [41] or the Finite-time rotation number [52]. See [47] for a review.

These indicators allow distinguishing among the ordered, chaotic or weak chaotic orbits, and even among the resonant and non-resonant regions [53]. However, Lyapunov exponents still remain valid indicators since they are quite easy to compute numerically, and they do not depend upon the metric. More importantly, in addition to mapping the "global" degree of instability, or presence of chaos, they can also reflect the local properties of the flow. This happens when they are computed during short intervals, including the time scales of regular-like behaviour of sticky chaotic orbits near remnants of periodic orbits embedded in the chaotic sea.

## 1.3   Computer Numerical Explorations

When discussing about the capability of calculating the future states of a system based on the past states, we have briefly mentioned the notion of computability. This concept deserves a deep discussion by itself, and the reader is pointed to, for instance, Ref. [35]. A computable operation, or equivalently, algorithmic, recursive or effective, can be defined as one operation that can be achieved by a Turing machine. There are alternative definitions based on the ideas of the American logician Alonzo Church, or those of the French mathematician Jacques Herbrand and the Austrian-American logician, mathematician, and philosopher Kurt Gödel, among others.

What is something remarkable is that there are some very well-defined mathematical operations that are definitively non-computable. But even constraining the discussion to the formally speaking computable problems, where it exist an algorithmic procedure for solving them, the resulting solution will not match with the real one. This is because there may be inherent differences between the actual real problem and the model used for making the predictions. But also because the numerical methods will introduce different errors and perturbations.

We must take into account these unavoidable sources of inaccuracies and errors, and face them from a practical point of view. If the system to be solved is only known approximately, it is meaningless to try to solve it with great accuracy. It can be enough to solve the approximate system and to focus on assuring that the numerical resolution will not introduce large errors, keeping them, at least, of the same order of magnitude than the error introduced by having an approximated model. But if the model does properly reflect the reality, or because any other reason, we must take into account the different sources of errors, and keep them under control. If this is not possible, due to the chaotic nature of the system, we should know at least what are the implications of their existence for the intended forecast.

### 1.3.1   Solving ODEs Numerically

The ordinary differential equations (ODEs) were born at the same time that the Infinitesimal Calculus, around the seventeenth century. Because the central role of these equations in the application of Mathematics to different sciences, Newton, Leibniz, Johann and Jacob Bernoulli make an effort in creating methods and techniques for solving them. Early attempts tried the elemental integration, based on changes of variable, algebraic manipulation, and problem-specific methods, for achieving the reduction to quadratures of a given problem. Soon, it was realised the difficulty of this task and more powerful schemes were introduced, such as the series methods, already used by Newton, and numerical methods [15, 16].

In general, a method is called *numerical* when it allows to obtain, even approximately, the solution to a mathematical problem by using a finite number of arithmetic operations: sums, subtractions, products and divisions. All numerical methods will discretise the differential system to produce a difference equation or map. Two aspects must be taken into account: how good will be the approximation to the solution, and the computational cost to reach it within a desired threshold.

As a very brief history on the subject, the first method to be born was raised by the Swiss mathematician Leonhard Euler in 1768 in his work "Institutionum Calculi Integralis, Volumen Primum". Euler's method is mainly used as the basis to construct more complex schemes with better error behaviour. But it can still be applied to very simple problems. Basically, it replaces the solution by a polygonal, Taylor expansion of first order. So, this method uses the present state of the system to provide the next state. Later on, Euler also invented other methods based on Taylor expansions of higher order.

The simplest Euler's method can be enhanced in several ways. The linear multi-step methods form one of these enhancements. The early linear multi-step methods were developed around 1850 by the British mathematician and astronomer John Couch Adams, and first appeared in the literature in 1885 [3]. These methods provide the solution in a future time using information of the recent past of the solution, instead of exclusively using the information from the present state, as the Euler's method does.

Another alternative way for enhancing Euler's method was introduced by the German mathematician Carl Runge in 1895 [40]. This was generalised later on by the German mathematician Karl Heun also [20], and, in special, by his fellow countryman Martin Wilhelm Kutta in 1901 [23]. In fact, Kutta introduced different notable methods, and one of them, one of the fourth order, and weights $1/6$, $2/6$, $2/6$ and $1/6$, is so frequently used that is sometimes known as the "Runge–Kutta method".

However, all that numerical calculus was made by hand, or at most, with the aid of mechanical devices in the early days. So, the practical use of numerical approaches for solving problems was not common. Nowadays, with the widespread use of computer simulations to solve complex dynamical systems, this scenario has dramatically changed. The numerical approaches are found today everywhere, and the reliability of those numerical calculations is of increasing interest.

### 1.3.2  Numerical Forecast

We can take a model, or set of equations describing the system, and integrate it during a certain time interval. The question to answer here is: How valid is the resulting forecast? Obviously, two initial points may diverge, or not, due to the presence of strong sensitivity to initial conditions. The larger this sensitivity, the larger the likelihood that a computed orbit will diverge from the real one.

Every model has inherent inaccuracies leading its results to deviate from the true solution. A given model can ignore certain variables, and may rely on certain assumptions and simplifications, making their results different from the solution. Even with a perfect model, the initial conditions may be based on experimental observations, meaning unavoidable experimental errors.

Another source of problems is formed by the computational issues. Computers will use float numbers, not real numbers, introducing also unavoidable errors. Finally, analytical solutions are not always possible, and numerical algorithmic computations must be done. Even the best method will diverge from the true orbit beyond certain time scales.

The numerical analysis aims at designing and building practical algorithms suitable for solving problems of mathematical analysis. Direct methods compute the solution to a problem in a finite number of steps, and may give the precise answer if they were performed in infinite precision arithmetic. Conversely, iterative methods start from an initial guess and converge to the exact solution by iterating successive approximations. In general, even using infinite precision arithmetic, these methods would not reach the solution within a finite number of steps. Therefore, a given tolerance residual decides when a sufficiently accurate solution has been found and the iterations must be stopped.

Whatever the method is selected, it will introduce several errors. Round-off errors are present because it is impossible to represent all real numbers exactly on a machine with finite memory. Truncation errors are committed when the iterative method is terminated or a mathematical procedure is approximated, and the approximate solution differs from the exact solution. Discretisation errors must be taken into account when the solution of the discrete problem does not coincide with the solution of the continuous problem. As a consequence, the selected algorithm should possess the property of keeping the resulting error without growing along the calculation. This is called numerical stability [38].

The stability concept is different than the accuracy, and tightly related to the discussion about as to whether two initially nearby orbits will remain in a defined neighbourhood or will move away from it, due to the errors, perturbations, or the noise introduced by the numerical scheme.

Independently of the numerical scheme or algorithm used for solving a given set of equations modeling a system, the system itself may present or not this stability property for propagating errors. So, independently on the source of this sensitivity, the problem to solve can be labeled as an "ill-conditioned" problem, due to the strong propagation of the initial errors. Otherwise, the problem is labeled as "well-conditioned".

Obviously, one would like to deal always with well-conditioned systems and well-conditioned algorithms. This is not always possible, but it can be somehow remediated by using a high-precision software. It uses efficient algorithms for performing, to any desired precision, the basic arithmetic operations, square and $n$-th roots, and even transcendental functions (an analytic function that does not satisfy a polynomial equation, in contrast to an algebraic function). There are several cases where this is of special interest, such as when solving ill-conditioned linear systems, large summations to execute on parallel computing, very long-time simulations, or even simple computations scaled up massively parallel systems. See [2] and the references therein for a review of these issues.

Of special importance is to use a stable method when solving a stiff equation. It seems that there is not a uniquely and accepted precise definition of stiffness, but it typically occurs when there are two or more very different scales of the independent variable on which the dependent variables are changing. When this happens, one needs to follow a variation in the solution on the shortest length scale to maintain the stability of the integration. The existence of two or more time scales in different directions of the dynamical flow, one quickly growing, one slowly growing, can lead to stiffness. In a stiff problem, the step size must be set to be extremely small even when apparently the solution is smooth. Otherwise the problem will be numerically unstable. The stiffness phenomenon is not a property of the exact solution but is a property of the differential system itself, because the same solution can be a solution of a set of stiff equations and a set of non-stiff equations [14, 24, 25].

We can classify the methods listed in the previous paragraphs to be explicit or implicit methods. The explicit methods are those when the new state is calculated from the current state. Typical examples are the forward Euler method, the midpoint method, second-order method with two stages, or "the Runge–Kutta method" of order 4. Because typically one may want an estimate of the local truncation error of a single Runge–Kutta step, there is a variety of methods called adaptive Runge–Kutta methods, intended to provide it. These include the Runge–Kutta–Fehlberg 54 method, or the Dormand–Prince DOP853, with step size control and dense output, among many others.

The implicit methods find a solution involving both the current and later state of the system. The implicit schemes usually require to solve a set of complex equations without analytical solution, so they are more complex and demand more computational resources. When the Runge–Kutta methods were performed manually, only low order schemes were used. Nowadays, there are explicit methods of order ten and implicit of arbitrary order [13]. When we find stiffness, implicit methods must be used. Otherwise, the use of an explicit method would require impractically small time steps to keep the errors within desired tolerance. For instance, regarding the Runge–Kutta methods, the stability function of an explicit Runge–Kutta method is a polynomial, so explicit Runge–Kutta methods can never be A-stable [21]. Because of their greater stability, implicit Runge–Kutta methods must be used when studying stiff equations.

From all the above, it is obvious the need of controlling the errors. We have seen the intrinsic difficulty of this task. This is of special relevance when there exist no suitable non-trivial exact solutions against which to test the results. There are some mitigating strategies, being the simplest one to run certain checks aiming to gain knowledge about the goodness of the computations [17]. The very first selection is to check the conservation of any integral of movement, if the system conserves it. The obvious choice is the total energy. But it may be of interest to monitor any other available integral of motion as the integration progresses.

Another cross-check to carry out can be to integrate time backwards the equations. This is not frequently done, and interesting conclusions can be obtained. The Burrau problem in the general three-body problem was reversed in [51], and the initial coordinates could be recovered under certain precision. And this leads to an interesting note when considering if checking the energy is enough as a claim about the goodness of the calculations. The recovery of the initial conditions was done to merely three significant figures even when the total energy was conserved to a relative accuracy much better than $10^{-11}$.

Ideally, another check is to perform independent calculations. Numerical explorations can be considered an experimental method, and, as such, should produce reproducible results. Different computations with the same initial conditions using different computers or different algorithms on the same computer should produce different results. In certain fields, this is not practical at all. We can take as an example the weather forecast. Here, the chaotic nature of the partial differential equations that govern the atmosphere joins to the required parametrisation for certain processes such as the solar radiation, precipitation and others, and it is impossible to solve these equations exactly.

For getting confidence in the numerical forecast in these cases, ensemble forecast techniques are routinely used for weather forecast. Ensemble forecasting can be seen as a form of Monte Carlo analysis, where different numerical predictions are done using slightly different initial conditions and different models or different formulations of a model. Ideally, the future dynamical system state should fall within the predicted ensemble spread, and the amount of spread should be related to the uncertainty of the forecast.

### 1.3.3 Symplectic Integrators

We have seen that it can be of interest to keep constant a given quantity along the integration, such as the energy in many cases. This raises the need of integrating the equations using symplectic schemes. These are numerical methods incorporating to the numerical schemes the geometric properties of the problem. This explains these techniques are also found under the label of "Geometric Integration".

From a mathematical point of view, Hamiltonian systems are just a specific set of systems within the broad world of possible differential equations. But they are almost everywhere in the field of applied mathematics, being present on all those systems without dissipation or with a dissipation that can be ignored, as it happens in key systems of Classical Mechanics, Quantum Mechanics, Optics, . . . One system is Hamiltonian if and only if the flow of their solutions is a symplectic transformation in the phase space. This notion of symplectic transformation is a geometric one. In cases with one degree-of-freedom, a symplectic transformation conserves the area of the flow. When there are higher number of degrees-of-freedom, the conserved area is multi-dimensional. The Poincaré's integral invariant is the most fundamental invariant in Hamiltonian Dynamics. For any phase space set, the sum of the areas of all of its orthogonal projections onto all the non-intersection canonically conjugate planes is invariant under the Hamiltonian evolution.

A Hamiltonian system integrated using an ordinary numerical method will become a dissipative, non-Hamiltonian, system with a completely different long-term behaviour, since dissipative systems do have attractors. When a Hamiltonian is integrated with a conventional method, the symplectic nature of the flow is not preserved and this is equivalent to modify, or perturb, the original system, and to solve a perturbed one, not being Hamiltonian any more.

Solving the system using a symplectic scheme can be seen as solving a perturbed system but without leaving the realm of Hamiltonian problems, because we will still be solving a perturbation of the real Hamiltonian function. This is important because Hamiltonian systems are not structurally stable against non-Hamiltonian perturbations, such as those introduced by classical explicit Runge–Kutta methods, independently of using fixed or variable step sizes. Some examples on how an explicit Runge–Kutta method of order 4, with variable step size (the standard ODE45) can deviate from real solution can be found in [33], for instance.

Historically, the first symplectic methods started from the idea that the symplectic transformations can be expressed in terms of a generation function, and tried to find a generation function suitable for numerical treatment. Another approach derived from showing that the Runge–Kutta methods can contain many symplectic methods [27, 42, 50]. It can be proven that all Runge–Kutta methods which are of the Gauss-Legendre type satisfy the necessary and sufficient condition for a Runge–Kutta method to be symplectic. Meanwhile, there is no explicit Runge–Kutta method that satisfies this condition, the simply-implicit methods can be made to obey such a condition. See [49] and the references therein for a review of all these issues.

There is an alternative to symplectic integrators, which are the exact energy-momentum conserving algorithms, designed to preserve these constant of motion. This scenario is not anyhow as simple as one may want. There is a theorem by Zhong and Marsden which says that there cannot exist integration schemes which are both symplectic and energy conserving for non-integrable systems: the only one to preserve both is the flow of the system itself [60]. This is impossible because the symplectic map with step $h$ would then have to be the exact time-$h$ map of the original Hamiltonian. Thus a symplectic map which only approximates a Hamiltonian cannot conserve the energy. So, one can find some algorithms which are energy conserving at the expense of not being symplectic.

As a rule of thumb, one may think that non-symplectic schemes behave well for small integrations, meanwhile symplectic schemes may be used for long integrations and a given variable must be conserved. That is, to consider that symplectic algorithms have generally a better performance in qualitative long term investigations, although the energy is only preserved on an average. In general, the lack of energy conservation is not a problem if the system is close to integrable, and has less than two degrees-of-freedom, because there will be invariant tori in the symplectic map which the orbit cannot cross, so the energy will remain oscillating, but bounded. Conversely, a non-symplectic method may allow the energy grow without limit.

But, in systems with higher number of degrees-of-freedom, symplectic integrations must be handled with care. Because meanwhile the original system may still have invariant tori, the scheme may add an extra degree-of-freedom, and this may lead to open holes on the system through Arnold diffusion [6].

Round-off errors are also a problem for Hamiltonian systems. They introduce again a non-Hamiltonian perturbation even if one uses a symplectic integrator. Even when such a perturbation could be less than the one seen before when using a non-symplectic integrator, it could still convert the system into a dissipative problem.

As a consequence, chaos itself can be indeed an artefact created from the numerical integration [19, 59]. And the use of symplectic integrators does not avoid a careful analysis of the system to solve. Classical high accuracy methods may be introduced for avoiding all the above-mentioned problems. Methods based on Taylor expansions work very fine in these cases, but the implementation needs to be carefully assessed on a problem-by-problem basis. See [11] and the references therein.

## 1.4  Shadowing and Predictability

Since the very beginning of Numerical Linear Algebra, it was of great interest to estimate the error when solving a set of linear equations. The first approach was the *progressive* analysis, that treated with the difference between the real solution and the computed solution. Later on, a second approach was attempted, the so-called *regressive* approach. By showing that the numerical solution exactly satisfies a system very close to the one that we are solving, one can try to enclose the difference between both systems. This is not just applicable when the progressive approach is not possible. In real cases, the system to solve is indeed only known approximately. For instance, the coefficients of the system may be subject to experimental errors inherent to any measurement. As a consequence, it can be considered meaningless to try to solve these approximate systems with great accuracy. Conversely, it may be enough to solve the approximate system just assuring that the numerical resolution will not introduce but similar errors to those found in the knowledge of the coefficients of the system.

The regressive approach interprets the calculated solution as a solution of a system that is a perturbation of the original problem to solve. This view gave birth to the concept of *shadowing*, that is the foundation of the concept of predictability as used along this monograph.

To label a system as *chaotic* means that due to the strong sensitivity to the initial conditions, two initial nearby trajectories will diverge with time, and at the same time, show complex behaviour. In mathematical terms, an irregular trajectory is defined as chaotic if it shows at least one positive Lyapunov exponent, the motion is bounded within certain limited region, and the $\omega$-limit set is not periodic neither composed of equilibrium points [1]. Conversely, a regular orbit will have vanishing Lyapunov exponents.

These chaotic motions can be analysed from the viewpoint of the Shannon theory of information. The entropy of a system, widely used in statistical mechanics, measures the probability of finding a system in a given state. The entropy is a measure of the disorder of the system, and using the Shannon theory of information, it gives the amount of information required for guessing a given state.

The computation of the Kolmogorov-Sinai entropy uses this idea for calculating the K-entropy as an average of the loss of information on the system states, calculated on all possible trajectories, per unit time. We can divide the phase space in different volume cells, and calculate the corresponding state at regular time intervals. Therefore, we can compute a sequence indicating in which cell the system was at a given time. The information given by the initial conditions will be lost once a given amount of finite time has elapsed. This interval will depend on the initial precision, and decreases as the K-entropy grows.

The K-entropy is closely related to the Lyapunov exponents values and the presence of algorithmic complexity found when the time sequence generated from one of its chaotic trajectories cannot be compressed by an arbitrary factor [11]. As a consequence, the loss of information in a given system is associated with the presence of positive exponents. If a system is chaotic, the reliability of the prediction is limited up to a time which is related to the largest Lyapunov exponent.

This time length, inverse of the asymptotic Lyapunov exponent, will be named here as *reliability* time. This time scale is also named as *Lyapunov time*, and is obviously related to the *e-folding* time, time interval in which an exponentially growing quantity increases by a factor of *e*. As alternate label, [17] considers the inverse of the finite-time Lyapunov exponent as *time scale of instability*.

One interesting finding is that there are real systems, such as chaotic asteroids, with Lyapunov times values that are rather small. This phenomenon was called "stable chaos" in [30]. Although those orbits have positive Lyapunov exponents, they are confined in thin chaotic layers for very long times, much longer than their respective Lyapunov times. These cases can be related to the stickiness property [54], where a chaotic orbit wanders close to the boundary of an island of stability, surrounded by a cantorus producing a barrier to the chaotic diffusion.

The term *predictability* of a system is understood along this book as a measure of the goodness of the system to make predictions, or forecasting, independently being chaotic or not. Chaos does not always imply a low predictability. An orbit can be chaotic and still be predictable, in the sense that the chaotic orbit is followed, or shadowed, by a real orbit, thus making its predictions physically valid. The computed orbit, also called pseudo-orbit, may lead to right predictions despite being chaotic because of the existence of a nearby exact solution. This true orbit is called a shadow, and the existence of shadow orbits is a very strong property.

The shadowing concept had a direct impact over the definition of the numerical methods [26, 43]. But the shadowing itself has a deeper impact on the dynamical systems to analyse. The real orbit is called a shadow, and the noisy solution can be considered an experimental observation of one exact trajectory. The distance to the shadow is then an observational error, and within this error, the observed dynamics can be considered reliable [44]. The shadows can exist, but it may happen that after

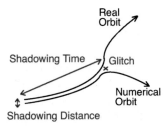

**Fig. 1.4** The shadowing time $\tau$ seen as the time a numerical trajectory keeps close to a true trajectory. The real orbit is called a *shadow*. The distance to the shadow is like an observational error, and within this error, the numerically observed dynamics can be considered reliable

a while, they may go far away from the true orbit. The real orbit is called a shadow, and the noisy computed solution can be considered an experimental observation of one exact trajectory. The distance to the shadow is then an observational error, and within this error, the observed dynamics can be considered reliable. The shadowing times define the duration over which there exists a model trajectory consistent with the real system and these shadowing times will be the basis to assess the predictability of our models (Fig. 1.4).

The shadowing can be found in hyperbolic dynamical systems, characterised by the presence of different expanding and contracting directions for the derivative. In hyperbolic systems, the angle between the stable and unstable manifolds is away from zero and the phase space is locally spanned by a fixed number of distinct stable and unstable directions [22, 56]. The shadowing can be found even in completely chaotic systems (Anosov systems), the strongest possible version of hyperbolicity where the asymptotic contraction and expansion rates are uniform (uniform hyperbolicity).

Non-hyperbolic behaviour can arise from tangencies (homoclinic tangencies) between stable and unstable manifolds, from unstable dimension variability or from both. In the case of tangencies, there is a higher, but still moderate obstacle to shadowing. But in other cases, the shadowing time can sometimes be very short, as, for instance, in the so-called pseudo-deterministic systems, where the shadowing is only valid during trajectories of given lengths due to the Unstable Dimension Variability (UDV).

Indeed, there are other aspects to take into account, and even regular orbits can also have shorter than expected predictability times. In strongly stiff systems, the predictability could be lower than expected for a regular and well-behaved, in appearance, orbit. In these systems, the existence of two or more time scales in different directions, one quickly growing, one slowly growing, can lead to shorten the time the computations can be physically meaningful.

## 1.5   Concluding Remarks

The presence of chaos is a concept different from the concept of stability of an orbit of a dynamical system. The stability characterises whether a perturbed orbit will remain in a neighbourhood of the unperturbed orbit or will go away from it. The presence of chaos means the presence of strong sensitivity to initial conditions, meaning that the future evolution of two initially close trajectories may be very different. We have reviewed some sources of errors. Because we deal either with experimental data or a numerical solution, and since even the best method will diverge from the true orbit beyond certain time scales, these errors are unavoidable when solving any physical model. So, even in the context of Classical Mechanics, the knowledge of future states of a deterministic equation seems to be limited, and beyond certain temporal boundary, the predictability time, the results from the computations can be fully unreliable. However, the shadowing theory provides some help here. The shadowing times define the duration over which there exists a model trajectory consistent with the real system and these shadowing times will be the basis to assess the predictability of our models.

## References

1. Alligood, K.T., Sauer, T.D., Yorke, J.A.: Chaos. An Introduction to Dynamical Systems. Springer, New York (1996)
2. Bailey, D.H., Barrio, R., Borwein, J.M.: High-precision computation: mathematical physics and dynamics. Appl. Math. Comput. **218**, 10106–10121 (2012)
3. Bashford, F.: An Attempt to Test the Theories of Capillary Action by Comparing the Theoretical and Measured Forms of Drops of Fluid with an Explanation of the Method of Integration Employed in the Tables Which Give the Theoretical Form of Such Drops, by J.C. Adams. Cambridge University Press, Cambridge (1883)
4. Carmona, R., Hwang, W., Torresani, B.: Wavelet analysis and its applications. In: Practical Time-Frequency Analysis: Continuous Wavelet and Gabor Transforms, with an Implementation in S, vol. 9. Academic, San Diego (1998)
5. Carpintero, D.D., Aguilar, L.A.: Orbit classification in arbitrary 2D and 3D potentials. Mon. Not. R. Astron. Soc. **298**, 21 (1998)
6. Cartwright, J.H.E., Oreste, P.: The dynamics of Runge–Kutta methods. Int. J. Bifurcation Chaos **2**, 427 (1992)
7. Chandre, C., Wiggins, S., Uzer, T.: Time-frequency analysis of chaotic systems. Physica D: Nonlinear Phenom. **181**, 171 (2003)
8. Cincotta, P.M., Simo, C.: Simple tools to study global dynamics in non-axisymmetric galactic potentials. Astron. Astrophys. **147**, 205 (2000)
9. Flaschka, H.: The toda lattice. II. Existence of integrals. Phys. Rev. B **9**, 1924 (1974)
10. Froeschlé, C., Lega, E.: On the structure of symplectic mappings. The fast Lyapunov indicator: a very sensitivity tool. Celest. Mech. Dyn. Astron. **78**, 167 (2000)
11. Gerlach, E., Skokos, Ch.: Comparing the efficiency of numerical techniques for the integration of variational equations. Discr. Cont. Dyn. Sys.-Supp. September, 475–484 (2011)
12. Gustavson, F.G.: On constructing formal integrals of a Hamiltonian system near an equilibrium point. Astron. J. **71**, 670 (1966)
13. Hairer, E.: A Runge-Kutta methods of order 10. J. Inst. Math. Appl. **21**, 47 (1978)

14. Hairer, E., Wanner, G.: Solving Ordinary Differential Equations II: Stiff and Differential-Algebraic Problems, 2nd edn. Springer, Berlin (1996). ISBN 978-3-540-60452-5
15. Hairer, E., Wanner, G.: Analysis by Its History. Springer, New York (1997)
16. Hairer, E., Norsett, S.P., Wanner, G.: Solving Ordinary Differential Equations, I, Nonstiff Problems, 2nd edn. Springer, Berlin (1993)
17. Heggie, D.C.: Chaos in the N-body problem of stellar dynamics. In: Roy, A.E. (ed.) Predictability, Stability and Chaos in N-Body Dynamical Systems. Plenum Press, New York (1991)
18. Hénon, M., Heiles, C.: The applicability of the third integral of motion: some numerical experiments. Astron. J. **69**, 73 (1964)
19. Herbst, B.M., Ablowitz, M.J.: Numerically induced chaos in the nonlinear Schroedinger equation. Phys. Rev. Lett. **62**, 2065 (1989)
20. Heun, K.: Neue Methode zur approximativen Integration der Differentialgleichungen einer unabhangigen Veranderlichen. Z. Math. Phys. **45**, 23 (1900)
21. Iserles, A.: A First Course in the Numerical Analysis of Differential Equations. Cambridge University Press, Cambridge (1996). ISBN 978-0-521-55655-2
22. Kantz, H., Grebogi, C., Prasad, A., Lai, Y.C., Sinde, E.: Unexpected robustness-against-noise of a class of nonhyperbolic chaotic attractors. Phys. Rev. E **65**, 026209 (2002)
23. Kutta, W.: Beitrag zur naherungweisen Integration totaler Differenialgleichungen. Z. Math. Phys. **46**, 435 (1901)
24. Lambert, J.D.: The initial value problem for ordinary differential equations. In: Jacobs, D. (ed.) The State of the Art in Numerical Analysis, pp. 451–501. Academic, New York (1977)
25. Lambert, J.D.: Numerical Methods for Ordinary Differential Systems. Wiley, New York (1992). ISBN 978-0-471-92990-1
26. Larsson, S., Sanz-Serna, J.M.: A shadowing result with applications to finite element approximation of reaction-diffusion equations. Math. Comput. **68**, 55 (1999)
27. Lasagni, F.M.: Canonical Runge-Kutta methods. Z. Angew. Math. Mech. **39**, 952 (1988)
28. Li, T., Yorke, J.A.: Period three implies chaos. Am. Math. Mon. **82**(10), 985 (1975)
29. Lorenz, E.: Deterministic nonperiodic flow. J. Atmos. Sci. **20**, 130 (1963)
30. Milani, A., Nobili, A.M., Knezevic, Z.: Stable chaos in the asteroid belt. Icarus **125**, 13 (1997)
31. Ott, W., Yorke, J.A.: When Lyapunov exponents fail to exist. Phys. Rev. E **78**, 056203 (2008)
32. Papaphilippou, Y., Laskar, J.: Global dynamics of triaxial galactic models through frequency map analysis. Astron. Astrophys. **329**, 451 (1998)
33. Pavani, R.: A two degrees-of-freedom hamiltonian model: an analytical and numerical study. In: Agarwal, R.P., Perera, K. (eds.) Proceedings of the Conference on Differential and Difference Equations and Applications, p. 905. Hindawi, New York (2006)
34. Penrose, R.: Quantum Implications. Essays in Honour of David Bohm. Routledge and Keegan P., London/New York (1987)
35. Penrose, R.: The Emperor's New Mind: Concerning Computers, Minds and the Laws of Physics. Oxford University Press, Oxford (1989)
36. Poincaré, H.: On the three-body problem and the equations of dynamics. Acta Math. **13**, 1 (1890)
37. Poincaré, H.: Les Méthodes nouvelles de la mécanique céleste. Gauthier-Villars et Fils, Paris (1892)
38. Press, W.H.: Numerical Recipes: The Art of Scientific Computing, 3rd edn. Cambridge University Press, Cambridge (2007). ISBN-10: 0521880688
39. Ruelle, D., Takens F.: On the nature of turbulence. Commun. Math. Phys. **20**, 167 (1971)
40. Runge, C.: Ueber die numerische Aflosung von Differentialgleichungen. Math. Anal. **46**, 167 (1895)
41. Sandor, Zs., Erdi, B., Szell, A., Funk, B.: The relative Lyapunov indicator. An efficient method of chaos detection. Celest. Mech. Dyn. Astron. **90**, 127 (2004)
42. Sanz-Serna, J.M.: Runge Kutta schemes for Hamiltonian systems. BIT **28**, 877 (1988)
43. Sanz-Serna, J.M., Larsson, S.: Shadows, chaos and saddles. Appl. Numer. Math. **13**, 449 (1991)
44. Sauer, T., Grebogi, C., Yorke, J.A.: How long do numerical chaotic solutions remain valid? Phys. Lett. A **79**, 59 (1997)

45. Simon, P., de Laplace, M.: A Philosophical Essay on Probabilities, p. 4. Wiley, New York; Chapman and Hall Ltd., London (1902)
46. Skokos, Ch.: Alignment indices: a new, simple method for determining the ordered or chaotic nature of orbits. J. Phys. A **34**, 10029 (2001)
47. Skokos, Ch.: The Lyapunov characteristic exponents and their computation. Lect. Notes Phys. **790**, 63 (2010)
48. Skokos, Ch., Bountis, T.C., Antonopoulos, Ch.: Geometrical properties of local dynamics in Hamiltonian systems: the Generalized Alignment Index (GALI) method. Physica D **231**, 30 (2007)
49. Stuchi, T.J.: Symplectic integrators revisited. Braz. J. Phys. **32**, 4 (2002)
50. Suris, Y.B.: Preservation of symplectic structure in the numerical solution of Hamiltonian systems. In: Filippov, S.S. (ed.) Numerical Solution of Differential Equations. Akad. Nauk. SSSR, p. 148. Inst. Prikl. Mat., Moscow (1988)
51. Szebeheley, V.G., Peters, C.F.: Complete solution of a general problem of three bodies. Astron. J. **72**, 876 (1967)
52. Szezech Jr., J.D., Schelin, A.B., Caldas, I.L., Lopes, S.R., Morrison, P.J., Viana, R.L.: Finite-time rotation number: a fast indicator for chaotic dynamical structures. Phys. Lett. A **377**, 452 (2013)
53. Tailleur, J., Kurchan, J.: Probing rare physical trajectories with Lyapunov weighted dynamics. Nature **3**, 203 (2007)
54. Tsiganis, K., Anastasiadis, A., Varvoglis, H.: Dimensionality differences between sticky and non-sticky chaotic trajectory segments in a 3D Hamiltonian system. Chaos Solitons Fractals **11**, 2281 (2000)
55. Valluri, M., Merrit, D.: Regular and chaotic dynamics of triaxial stellar systems. Astrophys. J. **506**, 686 (1998)
56. Viana, R.L., Grebogi, C.: Unstable dimension variability and synchronization of chaotic systems. Phys. Rev. E **62**, 462 (2000)
57. Voglis, N., Contopoulos, G., Efthymiopoulos, C.: Detection of ordered and chaotic orbits using the dynamical spectra. Celest. Mech. Dyn. Astron. **73**, 211 (1999)
58. Wellstead, P.E.: Introduction to Physical System Modelling. Academic, London (1979)
59. Wisdom, J., Holman, M.: Symplectic maps for the n-body problem, a stability analysis. Astron. J. **104**, 2022 (1992)
60. Zhong, G., Marsden, J.E.: Lie-Poisson Hamilton Jacobi theory and Lie-Poisson integrators. Phys. Lett. A **133**, 134 (1988)

# Chapter 2
# Lyapunov Exponents

## 2.1 Lyapunov Exponents

We have seen in the previous chapter that one of the fundamental questions about the dynamics of a system is to know if it is predictable or not. The answer to this question is tightly related to analyse if chaos is present in the dynamical flow.

Two trajectories starting out very close to each other may exponentially separate with time. In a finite time this separation may reach the scale of the accessible state space. In dissipative systems is of special interest to study these accessible spaces, and the different basin of attractions of the available attracting limit sets. All points in a neighbourhood of a trajectory may converge towards the same limit set, being a fixed point, a limit cycle, periodic or quasi-periodic motion, or a strange attractor. In conservative systems, where there are no attractors, one typically studies how to characterise these orbits that are constrained by the value of the conserved energy. Notably, the exponential divergence can be also found in these conservative systems.

Lyapunov exponents are a well-known diagnostic tool for analysing the presence of chaos, or chaoticity, of a system, and their utility comes in part from the fact that their values do not depend upon the metric. The ordinary, or asymptotic Lyapunov exponents describe the evolution in time of the distance between two nearly initial conditions, by averaging the exponential rate of divergence of the trajectories. They indicate the dynamical freedom of the system, because a larger exponent means a larger freedom, in the sense that small changes in the past lead to larger changes in the future.

The sensitivity to the initial conditions can be quantified by computing the exponential divergence or convergence of trajectories as follows:

$$\|\delta z(t)\| \approx e^{\lambda t}\|\delta z(0)\|, \tag{2.1}$$

---

The original version of this chapter was revised.
An erratum to this chapter can be found at DOI 10.1007/978-3-319-51893-0_5

© Springer International Publishing AG 2017
J.C. Vallejo, M.A.F. Sanjuan, *Predictability of Chaotic Dynamics*,
Springer Series in Synergetics, DOI 10.1007/978-3-319-51893-0_2

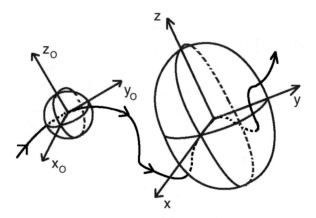

**Fig. 2.1** Evolution of an initial orthonormal basis centred around an initial point $x(0)$ after a given interval time has elapsed. The axes can be stretched and contracted, and their orientations can change too. The finite-time Lyapunov exponent $\chi(\Delta t)$ quantifies these changes averaging the rate of expansion or contraction during a time interval $\Delta t$. When this time interval tends to infinity, the finite-time exponents $\chi(\Delta t)$ tend to the asymptotic Lyapunov exponent $\lambda$

where $\lambda$ is a mean rate of separation of trajectories of the system, $\delta z(0)$ is the initial separation, and $\delta z(t)$ the separation after time $t$.

So, the Lyapunov exponent is defined in the following manner:

$$\lambda = \lim_{t \to \infty} \lim_{\delta z(0) \to 0} \frac{1}{t} \ln \frac{\delta z(t)}{\delta z(0)}. \tag{2.2}$$

One important remark to raise here is that the definition is based on the notion of distance between two phase space points. The Euclidean distance seems to be a natural choice in a continuous system, but other options may be applicable in other cases.

When the phase space is multi-dimensional, there can be different expansion rates along the different directions of the flow. So, in an $N$-dimensional flow there are $N$ different Lyapunov exponents, each one reflecting the averaged expansion rate of phase space in one direction along a given trajectory. If we consider a small hyper-sphere $N$-dimensional centred around the starting point or initial condition, the effect of the dynamics as the time elapses will be to distort the hyper-sphere into a hyper-ellipsoid, stretching its axes along some directions, contracting them along other directions (Fig. 2.1).

The ellipsoid axes will be distorted by the flow dynamics. The slopes of the flow in each direction provides a mean to know how the perturbation will evolve. The matrix describing these slopes is the Jacobian matrix of the flow, $J^t$, that describes these deformations after a finite time $t$. By solving at the same time the flow equation and the fundamental equation of the flow (that is, the distortion tensor evolution, see Appendix A for details), we can follow the evolution of the axes along the trajectory, and in turn, their growth rate.

The eigenvectors and eigenvalues are suited to study the iterated forms of a matrix, such as the Jacobian matrix. But the stretches are not related to the Jacobian matrix eigenvalues in a simple way, and the eigenvectors of the strain tensor $J^T J$ that determine the orientation of the principal axes are distinct from the Jacobian matrix eigenvectors. This is because the strain tensor does not satisfy any multiplicative property, and the principal axes must be recomputed from scratch for each time $t$.

When the eigenvalue of the system matrix is large, the errors in the initial condition are amplified by such a factor at local scales. At local dynamics level, knowing about the eigenvalues can help in selecting the most adequate method of integration. When the eigenvalue of the system matrix is very large, the error in the initial condition or the machine truncation can be amplified by that value.

The Lyapunov exponents can be seen as a generalisation of the eigenvalues at global scales. They indicate how much a given orbit diverge at global scale, by averaging the expansion rate of the phase space, and this average is needed to do because the expansion rate and the expansion direction change continuously. So, for a dynamical continuous flow, the asymptotic Lyapunov exponent can be defined as,

$$\lambda(x, v) = \lim_{t \to \infty} \frac{1}{t} \ln \|D\phi(x, t)v\|, \qquad (2.3)$$

provided this limit exists [48]. Here, $\phi(x, t)$ denotes the solution of the flow equation, such that $\phi(x_0, 0) = x_0$, and $D$ means the spatial derivative in the direction of an infinitesimal displacement $v$.

We should note that this equation implies a dependency on the direction of the distortion, that will disappear once the distortion is averaged and the dynamics tend to the fastest globally growing direction. However, since in practice the calculation is performed numerically, only a finite integration time is used instead of the infinite time defined in the equation above. So, an approximated value instead of the real one is returned. This is of course more important when working with experimental data, because of the very small number of measurements.

## 2.2 The Lyapunov Spectrum

The Lyapunov spectrum of an $N$-dimensional system is defined as the ordered set formed by the $N$ possible Lyapunov exponents. For $N$-dimensional flows, it is possible to have $N$ global Lyapunov exponents when a distortion tensor formed from $N$ perturbation vectors evolves according to the flow equations.

For a bounded orbit of an autonomous flow there is always an exponent with zero value in the limiting case (otherwise the system has an equilibrium in its limit set), as is tangent to the trajectory, and there is never any divergence for a perturbed trajectory in its direction [8].

**Table 2.1** Characterisation of attractors of a given flow dynamics based on the values of the Lyapunov spectrum

| Dimension | Attractor | $\lambda$ spectrum |
|---|---|---|
| 1 | Fixed point | $-$ |
| 2 | Periodic motion | $0\,-$ |
| 3 | Equilibrium point | $-\,-\,-$ |
| 3 | Limit cycle | $0\,-\,-$ |
| 3 | Torus $T^2$ | $0\,0\,-$ |
| 3 | Chaotic attractor $C^1$ | $+\,0\,-$ |
| 4 | Hypertorus $T^3$ | $0\,0\,0\,-$ |
| 4 | Chaos in $T^3$ | $+\,0\,0\,-$ |
| 4 | Hyperchaos $C^2$ | $+\,+\,0\,-$ |

This table is an adaptation of the one found in [34, 56]

Obviously, for dissipative flows the sum of all $\lambda$ exponents must be negative. Because there is always a zero value, in a two-dimensional flow, the first exponent is zero, and the second one must be negative. So, chaos is not possible in these flows, in agreement with the Poincaré-Bendixon theorem, stating that the limit sets are equilibria or periodic orbits.

Conversely, in higher-dimensional systems, there are more possible combinations of $\lambda$ values, and the set of all available exponents values is a convenient way for categorising asymptotic behaviours of dynamical systems. In this way, we can form a table like Table 2.1, adapted from Refs. [34] and [56]. Tables like this one summarise the possible combinations of the Lyapunov spectrum in dissipative systems.

The presence of chaos is given when there is at least one $\lambda$ exponent larger than zero. The regime is labeled as chaotic when only one global Lyapunov exponent $\lambda$ is positive. We talk about a "hyperchaotic" regime, also named sometimes as high dimensional chaos, when more than one positive Lyapunov exponent is present. This phenomenon is very important as is sometimes related to the presence of Unstable Dimension Variability (UDV) [64], as we will discuss further.

The sum of all Lyapunov exponents can be seen as the divergence of the flow, indicator of the overall expansion or contraction of the phase volume $V(t)$. Regarding the trace of the Jacobian, one can write:

$$\dot{V}(t) = \text{Tr}(J(t))V(t), \tag{2.4}$$

$$\frac{dV(t)}{V} = \text{Tr}(J(t)), \tag{2.5}$$

$$V(t) = e^{\int_0^t (\text{Tr}J(t))dt}. \tag{2.6}$$

For conservative $N$-dimensional systems, this sum must be zero. Because of the symplectic nature of the flow, the exponents fulfill the property of being in inverse additive pairs $\lambda_i = -\lambda_{N-i}$. In two-dimensional conservative systems, the two exponents are zero, and the motion is quasi-periodic and confined to a torus. But in higher dimensions, chaos is possible.

Regular quasi-periodic motions imply that all Lyapunov exponents are zero. But this can also happen in dynamical systems showing irregular dynamics where the separation of nearby trajectories grows weaker than exponential, implying zero Lyapunov exponents. These systems are sometimes referred to as weakly chaotic systems [33]. It is worthy to note that the term weak chaos can have different meanings. Some papers use this term when there is a smaller Lyapunov exponent, or two sets of trend values [44]. Others use it when the phase space dynamics is mainly regular with just a few chaotic trajectories and the dynamics is strongly dependent, in a very complex way, on the chosen initial condition [15]. In these weakly chaotic systems, there is no equivalence between time and ensemble averages. This weak ergodicity breaking means that a random sampling of the invariant distribution should not have the same content statistically speaking as a single orbit integrated for extremely long times.

We note here that analysing whether the Lyapunov exponents are zero or not can be useful for distinguishing between chaotic or non-chaotic orbits, but not for distinguishing irregular non-chaotic orbits. An irregular motion is chaotic if it is bounded, the $\omega$-limit set does not merely consist in connecting arcs and there is at least one asymptotic positive Lyapunov exponent [2]. Conversely, a regular orbit has vanishing Lyapunov exponents. However, it is not clear whether an irregular orbit will necessarily have at least a non-zero real exponent. Although it is generally assumed that irregular orbits and chaotic orbits are the same in Hamiltonian systems, this has not been proven in general [11].

Coming back to dissipative systems, the dimension of the existing attractors can be a useful indicator of its complexity. There is a variety of definitions for the term dimension. But in what concerns us, we will focus on the so-called Kaplan–Yorke dimension, $D_{YK}$, or "Lyapunov Dimension".

Kaplan and Yorke defined this dimension based on the value of the Lyapunov exponents [32]. They sorted all $\lambda$ values of the $N$-dimensional flow. Considering $D$ as the maximum number where $\lambda_1 + \lambda_2 + \ldots \lambda_D > 0$, one can define a topological dimension $D$, and one can consider that the dimension of the attractor lies between $D$ y $D + 1$.

The Lyapunov dimension, or Kaplan–Yorke dimension, is then defined as,

$$D_{KY} = D + (\lambda_1 + \lambda_2 + \cdots + \lambda_D)/|\lambda_{D+1}|. \qquad (2.7)$$

The Kaplan–Yorke conjecture is that this dimension is the same that the dimension of the attractor, and equal to the dimension of information. Any chaotic flow will have $D_{KY} > 2$, and chaotic maps can have any dimension. Going further, the dimension of the Lyapunov spectrum can be used as a valid estimator of the production of entropy in a system. The term $D_{KY}$ can be seen then as an upper limit to the dimension of information of the system, and following the theorem of Pesin, the sum of all positive exponents will return an estimation of the Kolmogorov–Sinai entropy [52].

As a final remark, when dealing with systems with a very high number of dimensions, as coupled maps, oscillators chains or disordered systems, the definition of Lyapunov spectrum can be generalised and can be defined as the ordered sequence of maximal exponents $\lambda_i$'s (in descending or ascending order) given as a function of $i/D$, where $D$ is the total number of exponents, equal to the dimension of the system. In these systems the existence of a limit spectrum when $D \to \infty$ [38] is assumed. It results rather interesting to study this spectrum. As a matter of fact, the properties of this spectrum have been studied in [49] starting from the properties of a infinite set of matrices.

## 2.3   The Lyapunov Exponents Family

It should be noted that neither the notation nor the definitions are standard in the literature. Since this can produce some confusion, it would be worthy to summarise some of the different notations found in the literature.

As we discussed before, the natural objects in dynamics are the linearised flow described by the Jacobian $J^t$, and its eigenvalues and eigenvectors, sometimes respectively referred to as stability multipliers and covariant vectors or covariant Lyapunov vectors [20].

Within this view, one can visualise the Lyapunov exponents as the normal modes of the system, and the solution of the flow, a linear combination of them. See [16] the and references therein. These eigenvalues of the Jacobian matrix are also sometimes called *local* Lyapunov exponents.

The Lyapunov exponents as asymptotic measures characterising the average rate of growth, or shrinking, of small perturbations to the solutions of a dynamical system were introduced by Lyapunov when studying the stability of nonstationary solutions of ordinary differential equations [39].

The existence of these exponents is assured when the general conditions of the Oseledec ergodic multiplicative theorem are fulfilled [46]. In this article, Oseledec refers to them as *Lyapunov characteristics numbers*, but Eckmann and Ruelle label them in [18] as *Characteristics exponents*. When the exponents exist, their values are independent from the initial condition. But these conditions may not always be met in common systems, as indicated in [48].

These exponents are generically labeled as *finite-time* Lyapunov exponents, independently of the finite interval length used in their computation. Notice that these exponents are sometimes named as *effective* Lyapunov exponent for large but finite intervals [25], meanwhile the term *local* Lyapunov exponent is preferred when such interval is small enough. The term *transient* Lyapunov exponent is found in [31], meaning intervals not large enough to ensure a satisfactory reduction of the fluctuations but small enough to reveal slow trends. Finally, the finite *size* Lyapunov exponents [10] analyse the growth of finite perturbations to a given trajectory, conversely to the analysis of the growth of infinitesimal perturbations done by the finite-time exponents. For our purposes, we will use the generic *finite-*

*time* Lyapunov exponents naming, independently of the length of the considered interval, and we will follow Eq. (2.11) for the exponents computations.

The term *finite-time* Lyapunov exponent (FTLE) is the one we have selected in this book. This term is also used for instance in [3, 45] and [55]. The notation of *direct* Lyapunov exponent (DLE) was used in [26], but this concept was a synonym of FTLE.

As mentioned in [25], the finite-time Lyapunov exponents are named as *effective* Lyapunov exponents when the intervals are considered large enough, but finite. Conversely, [19] and [31] prefer to use the term *transient* Lyapunov exponent when the finite interval is large enough for showing long time scales trends, but small enough for reducing short time scales fluctuations. A related label found in [17] is *Lyapunov Characteristic Indicator*, or LCI, which correlates with the previous definition of finite-time exponent.

The term *effective* Lyapunov exponent is also used in [13], meanwhile the term *local* Lyapunov exponent is used in [8] and [23]. These exponents are named as *localised* in [72]. This is because these authors refine this definition, and differentiate between *finite-time* and the *finite-sample* exponents. When calculating the exponents from an initial set of $N$-orthogonal vectors, corresponding to the initial deviation vectors, this set undergoes a few transient steps as their initial directions evolve under the system dynamics. After a few steps of integration and orthonormalisation, they could be considered already locally characteristic, meaning specific to a certain local flow. The *finite-time* refers to the case when the directions coincide with the right singular vectors of the matrix resulting from the Jacobian product, and the *finite-sample*, to the case when they correspond to the vectors resulting from the evolution of those singular vectors some steps before starting the computations. That is, when the growing rates are calculated using the direction of the Lyapunov vectors (i.e., the globally fastest growing direction as provided by the successive Gram–Schmidt orthonormalisations used by the algorithm of Benettin et al. [7]).

Another related term is the *averaged* finite-time Lyapunov exponent (AFTLE), used in [59], referring to finite-time exponents averaged on a large set of initial conditions. This is used for estimating the duration of chaotic transients.

Another widely found terminology derives from the *Lyapunov Number* (LN). This is defined as $\lambda = \log_e(\text{LN})$, as for instance in [2]. In general, the symbol $\lambda$ is used for the standard Lyapunov exponents, and we follow this trend. But in [68] this symbol is used for the Lyapunov Number and $\sigma$ is used instead for the standard exponent.

The LN is sometimes referred to as *Lyapunov Characteristic Number*, or LCN, and the Lyapunov Exponent as *Lyapunov Characteristic Exponent*, or LCE. In [5] and [60] it is also used the term *Maximal Lyapunov Exponent* (MLE) as an equivalent name to the LCE.

Some authors as [30] or [40] use a related name by defining *short-time* Lyapunov characteristic numbers as follows:

$$\chi(\Delta t) = \lim_{\delta z(0) \to 0} \frac{1}{\Delta t} \ln \frac{\delta z(\Delta t)}{\delta z(0)}, \qquad (2.8)$$

so the definition points to be the *finite-time* exponent.

We note here that the LCN term is also used in [12], but pointing to:

$$\chi = \frac{1}{t} \ln \frac{\delta z(t)}{\delta z(0)}. \qquad (2.9)$$

So, in this case, the LCN seems to be the LCE. This is also the case in [60]. This author specifies the LCN (i.e. the LCE) as,

$$\lambda = \lim_{t \to \infty} \chi(t), \qquad (2.10)$$

and the $\chi$ symbol is used for the so-called *Effective Lyapunov Number* (ELN). For making things easier, this author mentions that this ELN was named in the past *Local Characteristic Number*. This is the *effective* Lyapunov Exponent, and along this monograph, we have used the symbol $\chi$ when using finite-time calculations.

The term *Stretching Number* is used in [22] and [65] for conservative maps. This is the Lyapunov exponent calculated for inifinitesimal displacements and one iteration of the map. In [13], it is generalised to flows, and the interval is considered variable and equal to the integration step, a system characteristic time or a Poincaré section time, but in any case very small. Again, this seems to be the *finite-time* exponent.

Following this author, the mean (or first moment) of the distribution of stretching numbers is the LCN (i.e. LCE). This distribution of stretching numbers is named as *generalised Lyapunov indicator* in [22].

The term *finite-size* Lyapunov exponent, or FSLE, is used in [5] and [41]. These are averages of initial perturbations computed until a given error tolerance is reached. They measure the time that the perturbation grows a given factor at meso-scales. When both the perturbation and the error are infinitesimal, we have the standard exponent. In [24], a variant of the FSLE was introduced, as the *scale-dependent* Lyapunov Exponents, or SDLE. These are FSLE calculated using an algorithm optimised for very noisy time series.

When dealing with discrete-time systems, we find also the *finite-space* exponents, as in [35]. They measure the averaged divergence in discrete-time systems, tending to the standard exponent when the cardinality of the discrete phase space tends to infinity.

Finally, the term *conditional* Lyapunov exponent is used in [4] and [51] in the context of chaotic systems synchronisation. They are based on the idea that the conditions given by the Oseledec theorem assuring the existence of Lyapunov exponents also assure the existence of exponents in sub-blocks of the tangent map matrix. These sub-blocks exponents are, as consequence, indicators of the degree of freedom of the different system dynamics components.

For our purposes, we will use the generic *finite-time* Lyapunov exponents term, independently of the length of the considered interval, and we will follow Eq. (2.11) for the exponents computations.

In what concerns the Lyapunov spectrum, the distributions of finite-time Lyapunov exponents are sometimes referred to as Lyapunov spectra, but we will reserve such a term for the set of $N$ asymptotic values corresponding to an $N$-dimensional system, keeping in mind that one can say that every distribution of a single finite-exponent shows a range or "spectrum" of values.

The so-called Lyapunov vectors concept is linked to the Lyapunov exponents. The Lyapunov vectors are those vectors pointing in the direction in which the infinitesimal perturbation will grow asymptotically, exponentially at an average rate given by the Lyapunov exponents. In other words, a perturbation in the direction of a Lyapunov vector implies an asymptotic growth rate not smaller than $\lambda$ and almost all perturbations will asymptotically align with the vector pointing in the fastest growing direction. These vectors do not correspond to the eigenvectors of the Jacobian, or covariant Lyapunov vectors, which are a local entity and only require a local knowledge of the system.

The concept of bred, or breeding, vectors is tightly related to this. The bred vectors are created by adding random perturbations to an unperturbed initial condition, and both trajectories, the perturbed and the unperturbed, are subtracted from time to time as the integration takes place. This difference is the bred vector, and it must be scaled to be of the same size as the initial perturbation. Afterwards, this bred vector is added again to the unperturbed trajectory to create a new perturbed initial condition. Once this "breeding" process is iterated a few times, it will lead to bred vectors dominated by the fastest-growing instabilities [28, 29].

## 2.4   Finite-Time Exponents

The global Lyapunov exponents provide an indication on the globally averaged chaoticity of the system during an infinite integration time. But while they measure the asymptotic divergence of infinitesimally neighbouring trajectories, it is not always possible or desirable to perform these very long integrations and the limit value. Indeed, sometimes the asymptotic limit, thus the exponents themselves, may not exist [48].

In practice, all numerically computed exponents are computed over finite-time intervals. Such values are generically named as *Finite* Lyapunov exponents. Unlike the global Lyapunov exponents, which take the same values for almost every initial condition in every region if chaoticity is sufficiently strong (except for a Lebesgue measure zero set, following Oseledec theorem), the values of the exponents over finite times are generally different and may change in sign along one orbit.

We have seen the variety of notations and definitions regarding the finite-time Lyapunov exponents. For our purposes, we will focus on the following definition,

$$\chi(x, v, t) = \frac{1}{t} \ln \|D\phi(x, t)v\|, \tag{2.11}$$

which is derived from Eq. (2.3) for finite averaging times. Obviously, $\lambda = \chi(\Delta t \rightarrow \infty)$, with the implicit dependence on the point $x$ and the deviation vector $v$.

The finite-time Lyapunov exponents, computed according to this Eq. (2.11), reflect the growth rate of the orthogonal semiaxes (equivalent to the initial deviation vectors) of one ellipse centred at the initial position as the system evolves. Fixing this initial point, there are several choices for the initial orientation of the ellipse axes. Due to the dependence on the finite integration time interval used in Eq. (2.11), every orientation will lead to different exponents [72]. One option is to have the axes pointing to the local expanding/contracting directions, given by the eigenvectors, and at local time scales the eigenvalues will provide insight on the stability of the point. Other options are the axes pointing to the direction which may have grown the most under the linearised dynamics, or pointing to the globally fastest growing direction. In what concerns our technique, the initial axes of the ellipse are set coincident with a randomly set of orthogonal vectors, as in [62]. See Appendix A for details.

This option is selected because, as the flow evolves, the axes get orientated from the arbitrary direction as per the flow dynamics. The evolution in time of this orientation will depend on the selected finite-time interval length, that in turn will reflect the dynamical flow time-scales.

## 2.5   Distributions of Finite-Time Exponents

If we make a partition of the whole integration time along one orbit into a series of time intervals of size $\Delta t$, then it is possible to compute the finite-time Lyapunov exponents $\chi(\Delta t)$ for each interval. This section deals with the analysis of the evolution of the shapes of the finite-time Lyapunov exponent distributions as the finite-time interval size increases. This is because by doing so we can detect when the flow leaves the local regime and reaches the global regime [62].

We can get information about the degree of chaoticity of the orbit by subtracting different spectra [66], by deriving their power spectrum via the Fourier transform [67], or by analysing their shapes and cumulants or q-moments of the distribution. Such an approach has proved to be useful in several fields, such as galactic dynamics [50, 58], analysing chaotic fluid flows in the context of fast dynamos [21] or chaotic packet mixing and transport in wave systems[71].

The distributions of effective Lyapunov exponents can be studied from the cumulant generating function, defined as the logarithm of the moment generating function, which is itself the Fourier transform of the probability density function

[25]. The first four cumulants are the mean, variance, skewness and kurtosis of the distributions. As they reflect the deviation from Gaussianity, they reflect the deviation from the fully chaotic case. The *generalised* exponents are associated with the order-$q$ moments of the distributions [6, 14].

For some maps, like the Ulam map $x \mapsto 4x(1-x)$, explicit analytical expressions can be found to such probability exponents. In such cases, the probability distributions of time-$n$ exponents strongly deviate from the Gaussian shape, decaying with exponential tails and presenting $2^{n-1}$ spikes that narrow and accumulate close to the mean value with increasing $n$ [3]. Such tails and spikes were described for the Hénon–Heiles system in [61].

Here we aim to use the distributions for characterising different orbital behaviours. The shape of such distribution can serve as a valid chaoticity indicator, as it shows the range of values for $\chi$. In principle, the shape depends on the initial condition and on the sampling interval size $\Delta t$.

When considering the shapes of these distributions as a valid indicator, its evolution or stationarity is a key question. Here we follow some of the ideas started in [30, 40], where the dependence on the sampling time and the evolution towards an invariant measure in the distributions from orbits in chaotic domains were analysed. A clear description of how these spectra characterise the dynamical state in a set of hamiltonian prototypical cases was a motivation for our work.

Many distributions belonging to typical maps have been studied, as, for instance, in [17, 54, 64], but less consideration has been given to conservative systems, where no attractors are found. Indeed, we are interested in the distributions for characterising not only the possible final invariant measure, if so, but also the orbit itself, including the unstable and the open orbits (those that will escape towards the infinite). The main goal will be then to generate a set of protypical distributions for those different orbit behaviours.

The distribution of finite-time Lyapunov exponents can be normalised dividing it by the total number of intervals, thus obtaining a probability density function $P(\chi)$, that gives the probability of getting a given value $\chi$ between $[\chi, \chi + d\chi]$. Hence, the probability of getting a positive $\chi(\Delta t)$ or $P_+$ (and analogously $P_-$) can be defined as:

$$P_+ = \int_0^\infty P(\chi)d\chi. \tag{2.12}$$

Two ways for calculating such distributions are possible. The first one is starting from a given initial condition and integrating the flow equations during an interval $\Delta t$, therefore leading to have a $\chi(\Delta t)$ once the integration ends. Then, the integration starts again by taking the ending point of the previous integration as starting initial point of the new cycle. The second way is taking an ensemble of initial conditions on the available phase space (or energy surface). For each initial point, the value $\chi(\Delta t)$ is calculated as before, but without later progression in that orbit (see for instance [30, 40, 57]). When the phase space is largely chaotic and the regular regions small, both distributions coincide, in agreement with the ergodic theorem.

If the finite intervals are large enough, the expected shapes are a Gaussian, as the central limit theorem holds and the correlations die out. However, for small finite intervals, the shapes can be different. And when regular orbits appear, shapes can differ substantially.

So, we have computed the distribution by selecting an initial point of the orbit, an arbitrary set of orthonormal vectors, integrating then the flow up to $\Delta t$ interval, and resetting there the calculated effective exponent $\chi(\Delta t)$, starting again the integration in this new point.

Our first approach will be to compute the finite-time exponents using the smallest possible interval lengths. Later on, the interval size will be increased in order to see how the flow modifies the distributions.

Several criteria for choosing a small $\Delta t$ are found in the literature. The shortest interval that can be used in the case of maps is one iteration of the map. However for flows, as this time interval is a continuous quantity, several approaches are possible. It can be taken very small, although obviously it should not be smaller than the integration step. It has not been completely established yet whether these finite-time Lyapunov exponents distributions are typical or stationary when computed with short intervals $\Delta t$. See [53] for a discussion.

## 2.6   The Harmonic Oscillator

As a very simple initial example we will briefly analyse the behaviour of the Lyapunov exponents in the Harmonic Oscillator. We can take it as a representative example of the slow convergence rate of $\chi$ towards the asymptotic Lyapunov exponent $\lambda$.

The equations of the harmonic oscillator are the following:

$$\begin{cases} \dot{x} = p \\ \dot{p} = -\omega^2 x \end{cases} \tag{2.13}$$

where $\omega$ is the frequency of the oscillations. As arbitrary initial point, we can select $(-0.893978, 0.316862)$. We have also fixed $\omega^2 = 0.5$. Figure 2.2 plots the values of the finite-time Lyapunov exponents $\chi(\Delta t)$, as the finite-interval $\Delta t$ grows. We can see how the value slowly tends to the expected asymptotic value of $\lambda = 0.0$, but also we can see how this evolution is not monotonically decreasing even in this very simple case. Indeed, we see how the $\chi$ values are roughly constant for a while when the finite-intervals are very small.

The Poincaré section diagram for the harmonic oscillator is very simple, represented by a single plotted dot, with a crossing time between consecutive consequents $T_{\text{cross}} = T \sim 8.8$, being $T$ the period of the oscillations. Figure 2.2 also shows how the $\chi$ values tend to the final asymptotic good value $\lambda = 0$ when the finite-time length $\Delta t$ is larger than this period $T_{\text{cross}}$. So, we will analyse in the following sections the importance of this $T_{\text{cross}}$.

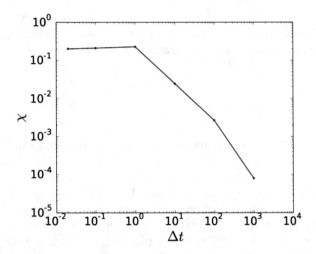

**Fig. 2.2** Evolution of the finite-time Lyapunov exponent $\chi(\Delta t)$ as the finite-time interval $\Delta t$ sizes are larger. For the largest lengths the values of the finite-time exponent $\chi(\Delta t)$ tend to reach the (asymptotic) Lyapunov exponent $\lambda = 0$

For the computation of the finite-time exponents, we will follow Appendix A. The flow of the harmonic oscillator system is given by,

$$\dot{v} = \Phi(v) = \begin{bmatrix} f_1(v) \\ f_2(v) \end{bmatrix} = \begin{bmatrix} p \\ -\omega^2 x \end{bmatrix} \tag{2.14}$$

By defining the Jacobian of the flow $\Phi$ as the matrix $J = D_v\Phi$, containing the differential slopes in every possible direction, we have:

$$D_v\Phi = \begin{bmatrix} \frac{\partial f_1}{\partial x} & \frac{\partial f_1}{\partial y} \\ \frac{\partial f_2}{\partial x} & \frac{\partial f_2}{\partial y} \end{bmatrix}, \tag{2.15}$$

In this specific simple, one-dimensional harmonic system, we have the following Jacobian,

$$D_v\Phi = \begin{bmatrix} 0 & 1 \\ -\omega^2 & 0 \end{bmatrix}, \tag{2.16}$$

The first issue is that the trace of the Jacobian is zero, as it should be. This reflects the conservative nature of the system. The second important point is that the Jacobian does not depend on the point of the trajectory, and, conversely, it is constant. So, regarding the distribution of values of $\chi(\Delta t)$, the value of $\chi(\Delta t)$

for a given $\Delta t$ and fixed initial condition is always constant along the orbit. As a consequence, the distribution of finite-time Lyapunov exponents for each $\Delta t$ size is a single $\delta$-Dirac, centred around the corresponding fixed finite-value $\chi(\Delta)$.

## 2.7  The Rössler System

Now, we will analyse a more complex system. The selected model consists of two identical, symmetrically diffusively coupled Rössler systems. We wish to describe its behaviour with the help of its global Lyapunov exponents. This system possesses a paradigmatic behaviour in relation to the chaos-hyperchaos transition and the Unstable Dimension Variability phenomenon, which was presented in [70] and [69] in a very similar system. In addition, it is a quite meaningful physical system, as it may represent the selective diffusion of two species through a semi permeable membrane in two continuously stirred tank reactors [42].

The equations of the system are

$$\begin{cases} \dot{x}_1 = -y_1 - z_1 \\ \dot{y}_1 = x_1 + ay_1 \\ \dot{z}_1 = b + z_1(x_1 - c) + d(z_2 - z_1) \\ \dot{x}_2 = -y_2 - z_2 \\ \dot{y}_2 = x_2 + ay_2 \\ \dot{z}_2 = b + z_2(x_2 - c) + d(z_1 - z_2) \end{cases} \qquad (2.17)$$

The first three coordinates $(x_1, y_1, z_1)$ correspond to the first Rössler oscillator. The second three coordinates $(x_2, y_2, z_2)$ to the other one. The parameter $d$ represents the coupling, which depends on the distance between the $z$-coordinates of the oscillators. The parameter $a$ is chosen as the control parameter. We have chosen the parameters $b = 2.0$ and $c = 4.0$, in order to compare our results with those from [69] and [70]. We have used a simple fourth-order Runge–Kutta method, with fixed time step 0.01 and a fourth-order/fifth-order Runge–Kutta–Fehlberg variable step size method as integrations schemes, both leading to the same numerical results. Figure 2.3 shows the two oscillators $(x_1, y_1)$ and $(x_2, y_2)$ for three different values of the control parameter $a$, and the final attractor of the oscillators. Note only that the $(x_1, y_1)$ and $(x_2, y_2)$ coordinates are displayed, being the $z$-component ignored. The plots are built with a total integration time of $T = 10{,}000$. The same initial condition is fixed in all cases to be $(1, 1, 0, -1, -5, 0)$, even when the same attractor is found when starting in the neighbourhood of this point. We see in this figure how the behaviour for both oscillators is different as the parameter $a$ changes.

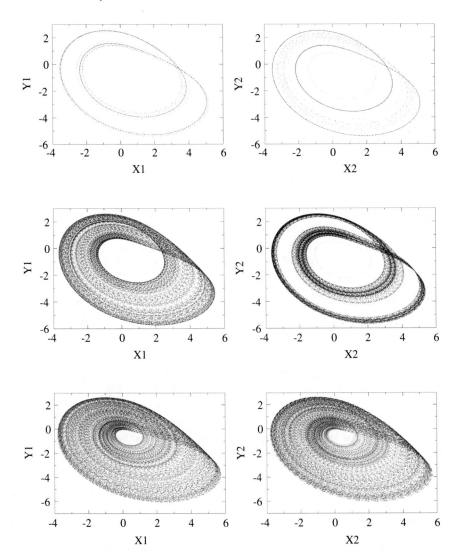

**Fig. 2.3** Evolution of two coupled Rössler oscillators $(x_1, y_1, z_1)$ and $(x_2, y_2, z_2)$ for three different values of the control parameter $a$. From top to above, $a = 0.342$, $a = 0.365$ and $a = 0.389$. These three cases are indicated in Fig. 2.7 as A, B and C. The figures show the values of $(x_1, y_1)$ and $(x_2, y_2)$. The coordinates $z_1$ and $z_2$ are not shown, for simplicity. Total integration time is $10,000$ time-units. A dot is plotted every 0.1 time-units. Coupling strength parameter is fixed as $d = 0.25$. Adapted from [63] with permission

Now, we will compute how the finite-time exponents $\chi(\Delta t)$ evolve as the intervals $\Delta t$ grow. Aiming to solve the variational equation, the Jacobian of the flow $J = D_v \Phi$, containing the differential slopes, is the following $6 \times 6$ matrix,

$$
D_v \Phi =
\begin{bmatrix}
\frac{\partial f_1}{\partial x_1} & \frac{\partial f_1}{\partial y_1} & \frac{\partial f_1}{\partial z_1} & \frac{\partial f_1}{\partial x_2} & \frac{\partial f_1}{\partial y_2} & \frac{\partial f_1}{\partial z_2} \\
\frac{\partial f_2}{\partial x_1} & \frac{\partial f_2}{\partial y_1} & \frac{\partial f_2}{\partial z_1} & \frac{\partial f_2}{\partial x_2} & \frac{\partial f_2}{\partial y_2} & \frac{\partial f_2}{\partial z_2} \\
\frac{\partial f_3}{\partial x_1} & \frac{\partial f_3}{\partial y_1} & \frac{\partial f_3}{\partial z_1} & \frac{\partial f_3}{\partial x_2} & \frac{\partial f_3}{\partial y_2} & \frac{\partial f_3}{\partial z_2} \\
\frac{\partial f_4}{\partial x_1} & \frac{\partial f_4}{\partial y_1} & \frac{\partial f_4}{\partial z_1} & \frac{\partial f_4}{\partial x_2} & \frac{\partial f_4}{\partial y_2} & \frac{\partial f_4}{\partial z_2} \\
\frac{\partial f_5}{\partial x_1} & \frac{\partial f_5}{\partial y_1} & \frac{\partial f_5}{\partial z_1} & \frac{\partial f_5}{\partial x_2} & \frac{\partial f_5}{\partial y_2} & \frac{\partial f_5}{\partial z_2} \\
\frac{\partial f_6}{\partial x_1} & \frac{\partial f_6}{\partial y_1} & \frac{\partial f_6}{\partial z_1} & \frac{\partial f_6}{\partial x_2} & \frac{\partial f_6}{\partial y_2} & \frac{\partial f_6}{\partial z_2}
\end{bmatrix} .
\tag{2.18}
$$

For this specific system the Jacobian is

$$
D_v \Phi =
\begin{bmatrix}
0 & -1 & -1 & 0 & 0 & 0 \\
1 & a & 0 & 0 & 0 & 0 \\
z & 0 & (x-c-d) & 0 & 0 & d \\
0 & 0 & 0 & 0 & -1 & -1 \\
0 & 0 & 0 & 1 & a & 0 \\
0 & 0 & d & r & 0 & (p-c-d)
\end{bmatrix} .
\tag{2.19}
$$

The trace of the Jacobian matrix is less than zero, indicating the dissipative nature of the system, and the existence of the attractors showed in Fig. 2.3. Figure 2.4 shows this evolution of $\chi(\Delta t)$ for the three different values of the control parameter $a$. Different regimes for three values of $a$ are reflected in the different convergence curves of the global Lyapunov exponents. We can see that the time required for reaching the asymptotic Lyapunov exponent is not the same in every case and depends on the value of $a$.

So, we have seen the time interval required for computing the asymptotic Lyapunov exponent $\lambda$, and we can consider $\chi(\Delta t = 100{,}000) \sim \lambda$. Now, we continue by calculating in detail how this asymptotic Lyapunov exponent depends on the control parameter $a$. We recall here that for $N$-dimensional flows, it is possible to have $N$ global Lyapunov exponents when a distortion tensor formed from $N$ perturbation vectors evolves according to the flow equations. When considering a single Rössler system, the first exponent can be just zero or positive, the second exponent is zero, and the third value negative, assuring the boundness of the solution. When two oscillators are coupled, a richer set of values is present. The chaotic regime is defined when only one global Lyapunov exponent $\lambda$ is positive, and the hyperchaotic regime, also named sometimes as high dimensional chaos, when there are more than one positive Lyapunov exponent.

The behaviour of the asymptotic exponents and raising of hyperchaotic transition, as parameters $a$ and $d$ are varied, is shown in Figs. 2.5 and 2.6. In the first figure, we fix the coupling parameter $d = 0.25$ and vary $a$. Below $a = 0.358$, all exponents are either nearly zero or below zero. Above this number, we have the chaotic regime,

**Fig. 2.4** Evolution of the
system of two coupled
Rössler oscillators $(x_1, y_1, z_1)$
and $(x_2, y_2, z_2)$. We see the
convergence towards the
global Lyapunov exponent of
the four largest finite-time
exponents from the total six
available exponents. The
remaining two exponents are
always negative and do not
provide additional
information, so they are not
displayed. The *upper row*
corresponds to $a = 0.342$,
Point A of Fig. 2.7. The
*middle row* to $a = 0.365$,
Point B of Fig. 2.7. The
*bottom* one $a = 0.389$, Point
C of Fig. 2.7. Adapted from
[63] with permission

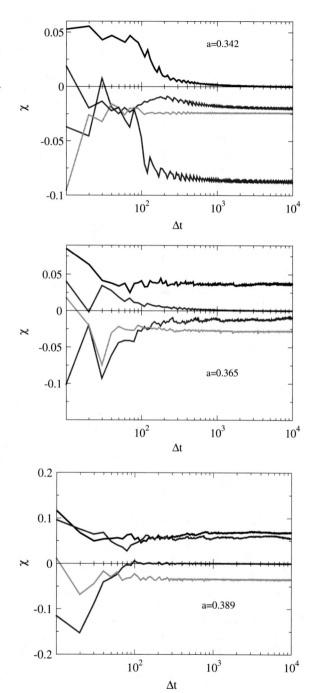

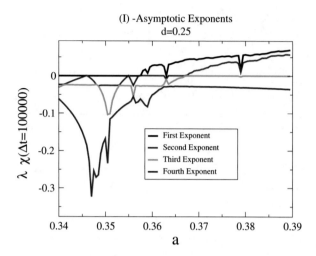

**Fig. 2.5** Lyapunov bifurcation diagrams or diagram showing the variation of $\lambda$ with the variation of the oscillator parameter $a$ and fixed coupling strength $d$. Hyperchaos is born at around $a \sim 0.367$. Only the four largest exponents from the total six are displayed. The remaining are always negative and are not shown. Global asymptotic Lyapunov exponents values $\lambda$ are calculated by computing $\chi(\Delta t = 100,000)$. Taken from [63] with permission

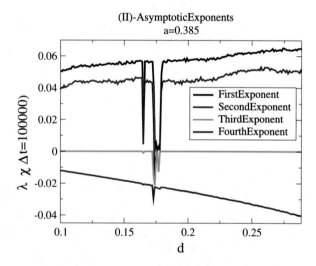

**Fig. 2.6** Lyapunov bifurcation diagram or diagram showing the variation of $\lambda$ with coupling strength $d$ and fixed parameter $a$. There is a drop in the hyperchaotic regime at $d \sim 0.174$. Only the four largest exponents from the total six are displayed. The remaining are always negative and are not shown. Global asymptotic Lyapunov exponents values $\lambda$ are calculated by computing $\chi(\Delta t = 100,000)$. Taken from [63] with permission

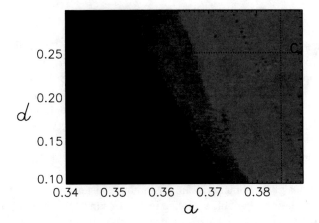

**Fig. 2.7** Hyperchaoticity chart. The number of positive global Lyapunov exponents varies with the Rössler parameter $a$ and the coupling strength parameter $d$. Dark regions (*Black and dark red*) mean 0 positive exponents (*dark red* meaning that the convergence is slower). Mid-bright regions (*red and dark pink*) mean only 1 positive exponent (*pink* meaning slower convergence). Brighter regions (*clear pink and above*) mean 2 positive exponents. *White* means 3. Slower convergence means that even with $\Delta t = 100,000$ the value has not reached zero limiting case within machine precision, but it is already smaller than $10^{-4}$. Points $A$, $B$, and $C$ are the three plots of Fig. 2.4. Slicing horizontally at $d = 0.25$ corresponds to Fig. 2.5 (top). Slicing vertically at $a = 0.385$ corresponds to Fig. 2.5 (bottom). Taken from [63] with permission

where there is at least one exponent larger than zero. From $a = 0.368$ there are at least two exponents, and the hyperchaotic regime starts. Note also that there is a window around $a = 0.381$ where both exponents decrease towards zero.

In the second figure, we fix $a = 0.358$ and vary $d$. For almost every coupling strength $d$, the system is hyperchaotic. However, there is a small interval around $d \sim 0.174$, where only the first global Lyapunov exponent remains positive. This shows that the chaos is not always decreasing (or increasing) with the coupling strength.

These different system regimes are displayed in Fig. 2.7, which shows the areas with no positive exponents (no chaos), just one positive exponent (chaos) and more than one positive exponent (hyperchaos). The hyperchaos arises in a complex way depending on parameters $a$ and $d$. There is no general trend of the hyperchaos with the coupling parameter $d$, as chaos sometimes increases and sometimes decreases with this parameter.

## 2.8   The Hénon–Heiles System

In the previous section we have seen the evolution of the finite-time Lyapunov exponents towards the asymptotic value, and how the spectrum of the finite-time Lyapunov exponents can trace the different regimes of the system. Now we are interested in analysing how the distributions calculated with the smallest available $\Delta t$ interval characterise a given system.

Even when some variability is expected when taken such small intervals, they can still serve for tracing the system. In fact, a way to determine the structure of a Lyapunov spectrum locally, that is, within some small (in principle infinitesimal) time interval is showed in [43]. Taking the interval size as small as possible, the correlation of each value to the following one will depend only on the local orbit behaviour. We will try to find out if the local information is enough for obtaining valid results or we should increase such interval.

When the interval size is increased, it can be made equal to any time interval with physical meaning, such as the characteristic time of the system or the crossing time of the orbit with a given Poincaré section. Moreover, instead of selecting a fixed $\Delta t$, it is possible to choose a variable sampling interval, as in [60], where it is taken to be equal to the interval where the $\chi(\Delta t)$ reaches a temporary limit. Anyway, one should keep in mind that when the size of $\Delta t$ is increased, the local details are washed out. And in the limit, $\chi(\Delta t \to \infty)$ tends to the asymptotic Lyapunov exponent, and the distribution tends to be a Dirac-$\delta$ centred at this asymptotic value.

We will review the behaviour of the finite-time Lyapunov exponents in the Hénon–Heiles system. The Hénon–Heiles system is one of these meridional potentials systems, and it was one of the first models used to show how a very simple system possesses highly complicated dynamics [27]. This model has been used as a paradigm in Hamiltonian nonlinear dynamics. It is a system with two degrees-of-freedom that in spite of its simplicity, it shows a very rich fractal structure in phase space when the system is open, that is, for energies above the threshold energy beyond which the orbits are unbounded.

Meridional plane potentials are those of the form $V(x, y) = V(R, z)$, being $R$ and $z$ the cylindrical coordinates, corresponding to an axisymmetric galaxy [9]. These are relatively simple potentials that can show complex behaviours, which are found in more realistic galactic-type potentials.

The motion in the meridional plane can be described by an effective potential:

$$V_{\text{eff}}(R, z) = V(R, z) + \frac{L_z^2}{2R^2}, \tag{2.20}$$

where $R$ and $z$ are the cylindrical coordinates. For each orbit, the energy $E = E(x, y, v_x, v_y)$ is an integral of motion. Once $E$ is fixed, only three of the four coordinates are independent and define the initial condition for the integrator.

If the energy $E$ and the $z$-component of the angular momentum $L_z$ are the only two isolating integrals, an orbit would visit all points within the zero-velocity curve, defined as $E = V_{\text{eff}}$. Sometimes, there are limiting surfaces that forbid the orbit to fill this volume, implying the existence of a third integral of motion, whose form cannot be explicitly written. In this case, the particle is confined to a $3 - torus$. Alternatively, there are some axisymmetric potentials where the orbits can indeed fill the meridional plane. These are irregular (or ergodic) orbits, which are only limited by two integrals of motion.

The Hénon–Heiles system was first studied by the astronomers Michel Hénon and Carl Heiles in 1964 [27], searching if there existed two or three constants of motion in the galactic dynamics. A system with a galactic potential that is axisymmetrical and time independent possesses a 6D phase space. As there are six variables, we can find five independent conservative integrals, some of them being isolating and other nonisolating (which are physically meaningless). The question that Hénon and Heiles tried to answer is which part of this 6D phase space is filled by the trajectories of a star after very long time. By that time, it was obvious that both the total energy $E_T$ and the $z$ component of the angular momentum $L_z$ were isolating integrals, while another two were usually nonisolating. Therefore, the real target became to find a third conserved quantity. In order to solve this problem, Hénon and Heiles proposed a 2D potential. Their result was that a third isolating integral may be found for only some few initial conditions. In fact, the Hénon–Heiles Hamiltonian is one of the first examples used to show how very simple systems might possess highly complicated dynamics, and since then, it has been extensively studied as a paradigm for two dimensional time-independent Hamiltonian systems. The Hénon–Heiles system can be seen as an approximation up to cubic terms to the Toda Chain, an integrable regular system typically modelling three atoms in a ring. This shows how an approximation to a regular system can exhibit chaos.

According to the energy of the orbit, which is related to the initial condition, different dynamical behaviours may appear and paradigmatic examples of the so-called pseudodeterministic models can be found. These models only yield relevant information over trajectories of reasonable length due to the unstable dimension variability [36, 37]. The oscillating behaviour of the finite-time Lyapunov exponents about zero has been found to be associated with these models [64].

The Hénon–Heiles Hamiltonian contains two, properly weighted, coupling terms, $x^2y$ and $y^3$, leading to a Hamiltonian with a $2\pi/3$ rotation symmetry and three exits in the potential well. It is written as

$$H = \frac{1}{2}(p_x^2 + p_y^2) + \frac{1}{2}\left(x^2 + y^2 + 2x^2y - \frac{2}{3}y^3\right). \tag{2.21}$$

So, as a consequence,

$$\begin{cases} \dot{x} = p \\ \dot{y} = q \\ \dot{p} = -x - 2.0xy \\ \dot{q} = -y - x^2 + y^2 \end{cases} \tag{2.22}$$

This Hamiltonian has been extensively studied for the range of energy values below the escape energy, where orbits are bounded and a variety of chaotic and periodic motions exist. On the other hand, if the energy is higher than this threshold value, the escape energy $E_e$, the trajectories may escape from the bounded region and go on to infinity through three different exits. The exit basins, understood as the set of initial conditions that lead to every individual exit, have a fairly complex structure fulfilling the Wada property [1].

As we are dealing with a two-degrees-of-freedom system, four Lyapunov exponents will exist. However, since it is a conservative Hamiltonian system, $\lambda_i = -\lambda_{5-i}$ for $(i = 1, \ldots, 4)$ and only two different values of $\lambda$ are independent. One of them will be tangent to the trajectory, parallel to the velocity field, and the other one, transverse to it. The tangent one is non-relevant as it tends to zero in the limit case.

The distribution of the finite-time or local Lyapunov exponents can be carried out by using standard methods. The initial ellipse axes are chosen arbitrarily. We can use a sixth-order Runge–Kutta integrator with a fixed time step equal to $10^{-2}$, because it provides enough accuracy for our purposes. The property $\lambda_i = -\lambda_{5-i}$ must be checked to be kept as the integration was evolving in time, to assure the goodness of the numerical computations.

The Poincaré section with the plane $x = 0$ has been plotted for each state, in order to compare the distribution with the dynamical state. We have selected this plane because of the symmetry of the system with respect to it, so each orbit must repeatedly intersect it. The crossing time $T_{\text{cross}}$ is defined as the time between successive section crosses, independently of the sign of $v_x$ when the plane $x = 0$ is crossed.

We have selected four initial conditions leading to four protypical behaviours in the Hénon–Heiles system. These orbits are listed in Table 2.2. The distributions of the first finite-time exponent $\chi(\Delta t)$ will be calculated following the same techniques described in the previous sections. The Jacobian of the flow $J = D_v \Phi$, containing the differential slopes, is the following $4 \times 4$ matrix,

$$D_v \Phi = \begin{bmatrix} \frac{\partial f_1}{\partial x} & \frac{\partial f_1}{\partial y} & \frac{\partial f_1}{\partial p} & \frac{\partial f_1}{\partial q} \\ \frac{\partial f_2}{\partial x} & \frac{\partial f_2}{\partial y} & \frac{\partial f_2}{\partial p} & \frac{\partial f_2}{\partial q} \\ \frac{\partial f_3}{\partial x} & \frac{\partial f_3}{\partial y} & \frac{\partial f_3}{\partial p} & \frac{\partial f_3}{\partial q} \\ \frac{\partial f_4}{\partial x} & \frac{\partial f_4}{\partial y} & \frac{\partial f_4}{\partial p} & \frac{\partial f_4}{\partial q} \end{bmatrix}. \tag{2.23}$$

For this specific system, the Jacobian is then as follows:

$$D_v \Phi = \begin{bmatrix} 0 & 0 & 1.0 & 0 \\ 0 & 0 & 0 & 1.0 \\ (-1.0 - 2.0y) & (-2.0x) & 0 & 0 \\ (-2.0x) & (-1.0 + 2.0y) & 0 & 0 \end{bmatrix}. \tag{2.24}$$

First, we raise a remark about periodic and quasi-periodic cases. In a periodic motion, as the harmonic oscillator, plotting the distribution of $\chi(\Delta t)$ values leads to a $\delta$-Dirac peaked distribution, as the $\chi(\Delta t)$ values repeat periodically and are constant. In cases where the Jacobian depends on the point, but the trajectory is still periodic, we will obtain Poincaré sections formed by closed curves, corresponding to the section of a torus. Regarding the shapes of the finite-time distributions, one typically obtains two-peaked distributions, corresponding to the regular motion.

**Table 2.2**  Selected orbits for the Hénon–Heiles system

| Orbit | Description | Initial condition for given energy | $\lambda$ | $T_{\mathrm{cross}}$ |
|---|---|---|---|---|
| H4 | Regular, close to period-5 orbit | $x = 0.000000 \; y = -0.031900 \; v_x = 0.307044 \; E = 1/8$ | 0.0 | 6.2 |
| H1 | Weakly chaotic, cycle orbit | $x = 0.000000 \; y = -0.119400 \; v_x = 0.388937 \; E = 1/12$ | 0.015 | 6.8 |
| H5 | Weakly chaotic, ergodic orbit | $x = 0.000000 \; y = -0.238800 \; v_x = 0.426750 \; E = 1/8$ | 0.044 | 6.8 |

$\lambda$ is the asymptotic standard Lyapunov exponent. The notion *weak* or *strong* chaos is associated with the relatively smaller or larger value of $\lambda$. $T_{\mathrm{cross}}$ is the Poincaré section crossing time corresponding to crosses with plane $x = 0$, independently of the sign of $v_x$

When the interval size increases, $\chi(\Delta t >>) \rightarrow 0$, and the peaks tend to merge and shift towards zero.

So, we start our analysis with a regular, quasi-periodic orbit, found in the Hénon–Heiles system for the energy $E = 1/8$. This is the orbit labeled as H4 in Table 2.2. The initial conditions are $x = 0.000000, y = -0.031900, \dot{x} = v_x = 0.307044$. Its Poincaré section is depicted in Fig. 2.8a, and it shows a set of ten islands, which is associated with a period-5 orbit. The five islands on the left are plotted when the $x = 0$ plane is crossed from the $x < 0$ subspace towards $x > 0$, and the other five on the right when returning to the $x < 0$ subspace. The distribution of finite-time Lyapunov exponents for an interval $\Delta t$ of 0.02 and total integration time of $10^4$ time units is the solid line in lower panel of Fig. 2.8b. It shows ten peaks, five centred around negative values and the other five centred around positive values.

In the inset panel, it has been plotted the evolution of the short time Lyapunov exponent with time (as the integrated number of intervals $\Delta t$ increases). As it is a quasi-periodic orbit, it can be observed quasi-periodic oscillations, with five oscillations per larger period. These oscillations in $\chi(\Delta t)$ are shown in the smaller inset panel. Each oscillation is associated to an island in the Poincaré section, thus to a peak in the distribution. Inside each period, we can count five oscillations, so five peaks are obtained in the distribution. Between each peak, a range of values is obtained, thus leading to the spectrum of values between the main peaks. As we are dealing with an orbit near a 5th-periodic one, only 10 peaks can be obtained. This means that there are arbitrarily finite intervals for which the orbit, on the average, is repelling in one of the dimensions and other intervals for which is attracting in the same dimension. The shape of the distribution is independent of the initial condition along the orbit, and longer integrations (in fact, larger than 600 time units, the circuit time or period to plot the five islands) do not lead to different shapes, because in this case we should be adding only more periods to the already sampled one. When the initial condition is moved far away from the periodic orbit, the distribution broadens but remains with a similar morphology.

When the interval size increases, the range of values around which the peaks are centred is reduced and it is shifted towards positive values, as shown in lower panel of Fig. 2.8b as dotted lines, and zoomed in the upper leftmost panel. When $\Delta t = 10$, a multipeaked distribution is still observed, since this value is larger than the crossing time but still smaller than the total circuit time, which is roughly 32 time units. This case is found as dashed-dotted in the upper rightmost panel. For larger size of time intervals the peaks begin to merge, as $\Delta t$ begins to be equal to that circuit time.

This behaviour is different for orbits showing some chaoticity. One example appears in Fig. 2.9a, with initial energy $E = 1/12$. This is the orbit labeled as H1 in Table 2.2. The initial conditions are $x = 0.000000, y = -0.119400, \dot{x} = v_x = 0.388937$. The solid line in Fig. 2.9b shows the corresponding distribution with an integration time of 20,000 units, and $\Delta t = 0.02$. The whole available phase space is traced and longer integrations lead basically to the same shape. This shape does not correspond to a"typical" chaotic state, where the central limit theorem holds for a number of averaged quantities, including local Lyapunov exponents (see [25] and

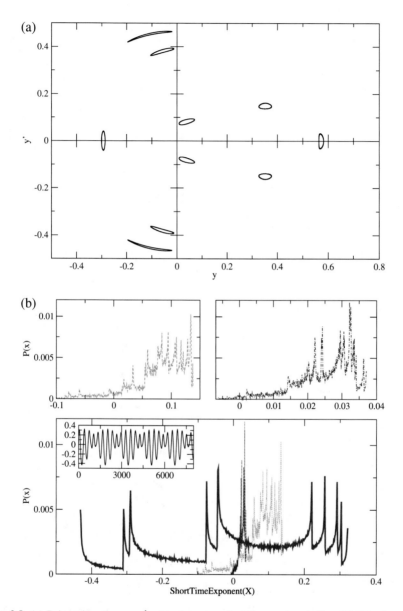

**Fig. 2.8** (a) Poincaré section $y - \dot{y}$ with plane $x = 0$ of the quasi-periodic orbit H4 of energy $E = 1/8$, associated with a period-5 orbit. The crossing time is $T_{\mathrm{cross}} \sim 6.2$ time units. Each time a point crosses the section, a different island is crossed and the total time before repeating an island is roughly 31.5 time units. (b) The *lower and larger panel* shows the distribution of finite-time Lyapunov exponents, showing 10 peaks both in positive and negative values, when $\Delta t = 0.02$ and total integration time $10^4$ time units. The *dashed* probability distribution is when $\Delta t = 10$ and integration time $10^6$, and is zoomed in the *upper leftmost panel*. The *dotted line* represents the probability distribution when $\Delta t = 100$ and integration time $10^6$, and is zoomed in the *upper rightmost panel*. Taken from [61] with permission

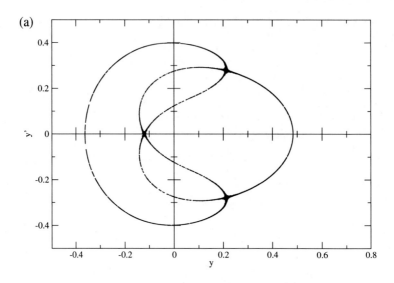

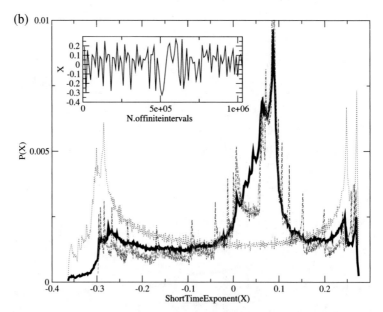

**Fig. 2.9** (**a**) Poincaré section $y - \dot{y}$ with plane $x = 0$ of the orbit H1 of energy $E = 1/12$. The crossing time is $T_{\text{cross}} \sim 6.8$ time-units. (**b**) The *solid line* shows the probability distribution of finite-time Lyapunov exponents formed with an integration of 20,000 time-units when $\Delta t = 0.02$. The *dotted and dashed lines* represent the probability distributions corresponding to partial 1000 time-units integrations started at arbitrary points of the same orbit. These partial integrations reflect some of the different transients of Table 2.3. The *smaller panel* shows the oscillating behaviour of $\chi(\Delta t)$ as the integration takes place. Taken from [61] with permission

**Table 2.3** Several distribution behaviours in the case $E = 1/12$ for the smallest interval size $\Delta t = 0.02$

| $t_0$ | Mean | Std. dev. | Median | $P_+(t_0)$ |
|---|---|---|---|---|
| 0 | −0.04402 | 0.18489 | −0.44017 | 0.43455 |
| $10^3$ | −0.01337 | 0.16457 | −0.01337 | 0.67674 |
| $2 \cdot 10^3$ | −0.01318 | 0.16437 | −0.01318 | 0.66708 |
| $3 \cdot 10^3$ | −0.01318 | 0.16438 | −0.01318 | 0.67882 |
| $12 \cdot 10^3$ | −0.04406 | 0.18492 | −0.04406 | 0.47806 |
| $14 \cdot 10^3$ | −0.016346 | 0.152195 | −0.016346 | 0.700080 |

The statistics are for integrations of $10^3$ time units starting at $t_0$

[47]) and the distributions can be fitted by a Gaussian, since the correlations die out. Neither does it to an intermittent system, where the shape might be a combination of a normal density and a stretched exponential tail, due to the long correlation persistence.

As we are analysing the evolution or stationarity of the distributions, it is important to keep in mind the difference between stationarity, due to the dynamics at certain time, and ergodicity, time-averaged property of the trajectories. In a non-ergodic orbit, the trajectory does not cover the whole hypersurface of constant energy, so two different initial conditions cover different parts of the energy surface leading to different temporal averages even for times tending to infinite. In such systems there is not a unique equilibrium state, but different ones depending on the starting point. Conversely, in an ergodic system it can be reached a unique equilibrium state. And generic ensembles of initial conditions will evolve towards a given distribution, time-independent or with little variability on long time-scales. One key point is the time involved in such evolution towards the final state. If the physical time scales are relevant and that time is too long for being realistic, those ensembles will not be able to be used as valid skeleton for the observed system behaviour.

So, when computing distributions from a set of initial conditions, we need to be sure they are in the same domain of the Poincaré section. If that is the case, we get again the solid histogram of Fig. 2.9b. On the other hand, the stationarity of a distribution can be defined when the statistical parameters do not change with time, and this depends on the variable dynamics along the given orbit. When the probability distribution from a single orbit is computed, the morphology may depend on the initial point, when the total integration time is not large enough, as several transients of different behaviour are found (see [72]).

In order to catch the behaviour of the transient periods, we have computed distributions formed after integrating just $10^3$ time units (150-times the crossing time), which are described in Table 2.3. Three of them appear in Fig. 2.9b. The characteristic time on which the orbit forgets its previous degree of instability is small (low correlation time), as they are quite different. The standard deviation of the distributions $\sigma$ gives a measure of the degree in which $\chi$ deviates from the mean, being a measure of the stability or variability of the values of $\chi$ along the orbit. The Eq. (2.12) returns $P_+$, which is the probability of getting a positive finite-time

**Fig. 2.10** The distribution of finite-time Lyapunov exponents in the case $E = 1/12$ formed with an integration of $10^6$ time units when $\Delta t = 1$ is plotted as solid trace in the *lower panel*. The same when $\Delta t = 10$ appears in *dashed line*, and is zoomed in the *upper panel*. In this latter one, is is also traced the distribution when $\Delta t = 100$. Taken from [61] with permission

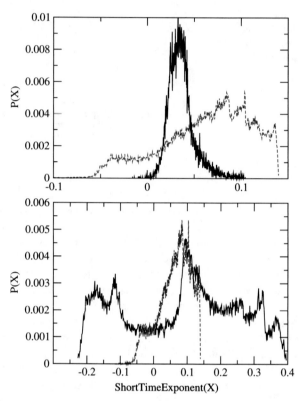

Lyapunov exponent. The probability $P_+$ takes different values ranging from 0.4 up to 0.7 quite randomly, what indicates different behaviours, ordered at some stages, chaotic in others, as reflected in the shape of the distributions. For instance, the first transient shows two well-separated peaks, like a quasi-periodic orbit (dotted line), while the third transient shows a multi-peaked distribution (dotted-dashed line). When the time evolution of the finite-time Lyapunov distributions (and the time evolution of $\chi(\Delta t)$ itself, as shown in the smaller panel) is compared with the way the consequents of the Poincaré section fill the available phase space, we see how each distribution corresponds to a different way of tracing the Poincaré section.

If we change the interval size $\Delta t$ by a small integer factor, our result is only a re-scaling of the spectrum, as was shown in [58]. However when it is increased up to, say, $\Delta t = 1$, which is still smaller than the averaged crossing time, a different multi-peaked shape is obtained, as shows the solid line in lower panel of Fig. 2.10. The local details are washed up as the interval size is larger than the crossing time, so with a $\Delta t = 10$ (dotted line), the shape is again different. This distribution is zoomed in the upper panel and two smooth peaks well fitted by Gaussians, the main one centred around positive values are observed. For even larger values of $\Delta t = 100$, a single peak Gaussian distribution is found, as plotted as solid in the upper panel of Fig. 2.10, since the central limit theorem begins to hold.

**Table 2.4** Several
distribution behaviours in the
case $E = 1/12$ for interval
size $\Delta t = 1$

| $t_0$ | Mean | Std. dev. | Median | $P_+(t_0)$ |
|---|---|---|---|---|
| 0 | 0.07362 | 0.17391 | 0.07362 | 0.46000 |
| $10^3$ | 0.09552 | 0.17119 | 0.09552 | 0.74000 |
| $2 \cdot 10^3$ | 0.09293 | 0.17478 | 0.09293 | 0.72000 |
| $3 \cdot 10^3$ | 0.09117 | 0.17415 | 0.09117 | 0.73000 |
| $12 \cdot 10^3$ | 0.06484 | 0.16613 | 0.06484 | 0.51000 |
| $14 \cdot 10^3$ | 0.08573 | 0.14284 | 0.08573 | 0.75000 |

The statistics are for integrations of $10^3$ time units
starting at $t_0$

**Table 2.5** Sensitivity of the statistics of the finite-time Lyapunov distributions in the case $E = 1/12$ for several integration time and interval sizes

| $t$ (total time) | $\Delta t$ (time) | Mean | Std. dev. | Median | $P_+(t_0)$ |
|---|---|---|---|---|---|
| $2 \cdot 10^4$ | 0.02 | −0.04403 | 0.18490 | −0.04403 | 0.64226 |
| $2 \cdot 10^5$ | 0.02 | −0.04407 | 0.18492 | −0.04407 | 0.65772 |
| $2 \cdot 10^4$ | 1 | 0.08553 | 0.17873 | 0.08553 | 0.69000 |
| $2 \cdot 10^5$ | 1 | 0.08454 | 0.18004 | 0.08454 | 0.71060 |
| $2 \cdot 10^4$ | 10 | 0.03215 | 0.06258 | 0.03215 | 0.90400 |
| $2 \cdot 10^5$ | 10 | 0.02509 | 0.06671 | 0.02511 | 0.89565 |

Finally, for much larger values of $\Delta t = 100$, the distributions collapse to $\delta$-functions centred around the asymptotic Lyapunov value. In addition, the chaoticity indicators vary with the interval size. The values in Table 2.4 are calculated as in Table 2.3, so here it appears the statistics for the transients with an integration (sampling) of $\Delta t = 1$.

The mean value calculated with the larger interval on each transient is different to the calculated with the smallest interval. Moreover, the values of $P_+$ are larger, and for even larger interval sizes, the transients may vanish. However, it is remarkable that the evolution of $P_+$, which is an indicator of the local chaoticity, is similar in both cases. Table 2.5 shows how the total integration time for a given interval size is correlated with these indicators, showing that for the smallest interval we obtain similar results.

An explanation is that by integrating $2 \cdot 10^4$, we have already passed through all possible values of the oscillating short time Lyapunov exponent, so even increasing the total integration time up $2 \cdot 10^5$, the spectrum is basically the same. For larger intervals, the statistics is poorer, as the total number of intervals taken into account is smaller, but the same reasoning can be done. When $\Delta t = 1$, we are still getting almost the same pattern in the oscillations with $2 \cdot 10^4$ time units or $2 \cdot 10^5$, so the values are still quite similar. But with $\Delta t = 10$, the values are slightly different, as the pattern of the oscillations of the short time Lyapunov exponent is also slightly different.

Finally, the characterisation of the distributions corresponding to chaotic orbits is discussed. We take an orbit with an initial energy $E = 1/8$, that almost completely fills the available phase space, as shown by the Poincaré section in Fig. 2.11a. This is the orbit labeled as H5 in Table 2.2. The initial conditions are $x = 0.000000, y = -0.238800, \dot{x} = v_x = 0.426750$. The corresponding distribution is plotted as solid line in Fig. 2.11b. The smaller panel shows again the oscillations of $\chi(\Delta t)$ as the integration takes place.

The same probability distribution is obtained by integrating along a single initial condition or an ensemble of initial conditions, due to the ergodicity of the system. The shape reminds the one described for attractors in [36, 54], although the tail of the peak centred around positive values extends through negative values quite smoothly, instead of showing an exponential tail. Two different transients of $10^3$ time units are plotted as dotted and dashed lines in Fig. 2.11b. We also see that the sticky orbits, those that remain near a regular island for a long time, tend to have smaller exponents than the non-sticky orbits. During the sticky periods, when the orbit appears next to a quasi-periodic orbit torus, the distribution is clearly similar to a quasi-periodic case. However, in the chaotic regime the peaks are broadened. With larger intervals ($\Delta t = 10$) and integration times ($10^6$ time units), an almost Gaussian shaped distribution is obtained, centred around a positive value. This shows a morphology different from the $E = 1/12$ case, that did not reach such Gaussian form even when $\Delta t = 10$, meaning a different dynamics, which is also manifested by the time the distribution takes to its final state.

Such different morphology can be seen by comparing the solid lines of Figs. 2.9b and 2.11b. In the latter case, the peak is not so clear, and the distribution is smoother, indicating that there is no larger probability of getting a value over another one. In the prior case, there is a clear peak, indicating that there is high probability of getting the range of values on which the peak is constructed. So the latter case indicates that there is more "chaoticity" in the sense that there are no privileged values, as in the $E = 1/12$ case, so there is a larger ergodicity, in the sense that the orbit is able to reach with the same probability all the available phase space. However, it should be taken into account that during certain transient periods, the behaviour is equivalent to ordered motions, as during the sticky transients (double-peaked distributions).

## 2.9  Concluding Remarks

We have seen that, in addition to the standard calculation of the asymptotic Lyapunov exponent, we can extract more information about the dynamical system by calculating the distributions of finite-time Lyapunov exponents. We have analysed the information provided by these distributions of finite-time Lyapunov exponents. Several prototypical distribution morphologies have been plotted for different energy values.

Shapes well differentiated depending on the motion type, the interval size and the integration time have been found.

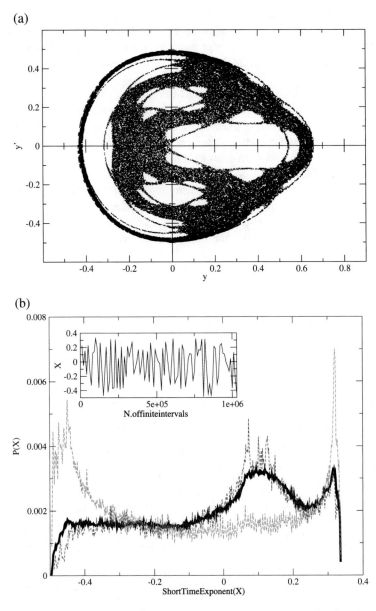

**Fig. 2.11** (a) Poincaré section $y - \dot{y}$ with plane $x = 0$ of the orbit H5 of energy $E = 1/8$. The crossing time is $T_{\text{cross}} \sim 6.8$ time units. (b) Probability distribution of finite-time Lyapunov exponents. The *solid line* corresponds to an integration of 20,000 time units when $\Delta t = 0.02$. The *dotted and dashed* ones to partial integrations of $10^3$ time units. The double peaked one corresponds to a sticky period. The *smaller panel* shows the oscillating behaviour of $\chi(\Delta t)$ as the integration takes place. Taken from [61] with permission

Our calculations have focused on the use of the smallest interval size, searching for the stationarity or evolution of the distributions. It has been observed that they characterise the motion in the different possible cases. The overall shape depends on the local orbit behaviour, as the exponents can be considered specific of a certain local flow.

We have detailed the results of our analysis on a Hamiltonian system. Here, the chaotic orbits are ergodic and the results from generating the distributions from an adequate ensemble or from a single orbit are equivalent. But we have also analysed the distributions of non-ergodic orbits. The results obtained with this approach should be valid for orbits both in conservative and in non-conservative systems.

The previous discussion shows some implications when the physical meaning of the system is taken into account. As the long integrations required for computing the asymptotic Lyapunov exponents may have no meaning, as for instance, in a galactic system, since the universe evolves in a shorter time, it is reasonable to use smaller integrations. Furthermore, the smallest interval sizes can be used since they characterise the local behaviour.

# References

1. Aguirre, J., Vallejo, J.C., Sanjuán, M.A.F.: Wada basins and chaotic invariant sets in the Hénon-Heiles system. Phys. Rev. E **64**, 66208 (2001)
2. Alligood, K.T., Sauer, T.D., Yorke, J.A.: Chaos. An Introduction to Dynamical Systems, p. 383. Springer, New York (1996)
3. Anteneodo, C.: Statistics of finite-time Lyapunov exponents in the Ulam map. Phys. Rev. E **69**, 016207 (2004)
4. Araujo, T., Mendes, R.V., Seixas, J.: A dynamical characterization of the small world phase. Phys. Lett. A **319**, 285 (2003)
5. Aurell, E., Boffeta, G., Crisanti, A., Paladin, G., Vulpiani, A.: Predictability in the large: an extension of the concept of Lyapunov exponent. J. Phys. A Math. Gen. **30**(1), 1–26 (1997)
6. Badii, R., Heinzelmann, K., Meier, P.F., Politi, A.: Correlation functions and generalized Lyapunov exponents. Phys. Rev. A **37**, 1323 (1988)
7. Benettin, G., Galgani, L., Giorgilli, A., Strelcyn, J.M.: Lyapunov characteristic exponents for smooth dynamical systems and for Hamiltonian systems; a method for computing all of them. Meccanica **9**, 20 (1980)
8. Benzi, R., Parisi, G., Vulpiani, A.: Characterisation of intermittency in chaotic systems. J. Phys. A **18**, 2157 (1985)
9. Binney, J., Tremaine, S.: Galactic Dynamics. Princeton University Press, Princeton, NJ (1987)
10. Boffetta, G., Cencini, M., Falcioni, M., Vulpiani, A.: Predictability: a way to characterize complexity. Phys. Rep. **356**, 367 (2002)
11. Carpintero, D.D., Aguilar, L.A.: Orbit classification in arbitrary 2D and 3D potentials. Mon. Not. R. Astron. Soc. **298**, 21 (1998)
12. Contopoulos, G., Voglis, N.: A fast method for distinguishing between ordered and chaotic orbits. Astron. Astrophys. **317**, 317 (1997)
13. Contopoulos, G., Grousousakou, E., Voglis, N.: Invariant spectra in Hamiltonian systems. Astron. Astrophys. **304**, 374 (1995)
14. Crisanti, A., Paladin, G., Vulpiani, A.: Product of Random Matrices. Springer Series in Solid State Sciences. Springer, Berlin (1993)

15. Custodio, M.S., Manchein, C., Beims, M.W.: Chaotic and Arnold stripes in weakly chaotic Hamiltonian systems. Chaos **22**, 026112 (2012)
16. Cvitanovic, P., Artuso, R., Mainieri, R., Tanner, G., Vattay, G., Whelan, N., Wirzba, A.: Chaos: Classical and Quantum. Niels Bohr Institute, Copenhagen (2016). ChaosBook.org
17. Diakonos, F.K., Pingel, D., Schmelcher, P.: analysing Lyapunov spectra of chaotic dynamical systems. Phys. Rev. E **62**, 4413 (2000)
18. Eckmann, J.P., Ruelle, D.: Ergodic theory of chaos and strange attractors. Rev. Mod. Phys. **57**, 617 (1985)
19. Ershov, S.V., Potapov, A.B.: On the nature of nonchaotic turbulence. Phys. Lett. A **167**, 60 (1992)
20. Ershov, S.V., Potapov, A.B.: On the concept of stationary Lyapunov basis. Physica D **118**, 167 (1998)
21. Finn, J.M., Hanson, J.D., Kan, I., Ott, E.: Steady fast dynamo flows. Phys. Fluids B **3**, 1250 (1991)
22. Froeschlé, C., Lohinger, E.: Generalized Lyapunov characteristic indicators and corresponding Kolmogorov like entropy of the standard mapping. Celest. Mech. Dyn. Astron. **56**, 307 (1993)
23. Fujisaka, H.: Statistical dynamics generated by fluctuations of local Lyapunov exponents. Prog. Theor. Phys. **70**, 1264 (1983)
24. Gao, J.B., Hu, J., Tung, W.W., Cao, Y.H.: Distinguishing chaos from noise by scale-dependent Lyapunov exponents. Phys. Rev. E **74**, 066204 (2006)
25. Grassberger, P., Badii, R., Politi, A.: Scaling laws for invariant measures on hyperbolic and non-hyperbolic attractors. J. Stat. Phys. **51**, 135 (1988)
26. Haller, G.: Distinguished material surfaces and coherent structures in 3d fluid flows. Physica D **149**, 248 (2001)
27. Hénon, M., Heiles, C.: The applicability of the third integral of motion: some numerical experiments. Astron. J. **69**, 73 (1964)
28. Kalnay, E.: Atmospheric Modeling, Data Assimilation and Predictability. Cambridge University Press, Cambridge (2007)
29. Kalnay, E., Corazza, M., Cai, M.: Are bred vectors the same as Lyapunov vectors? EGS XXVII General Assembly, Nice, 21–26 April 2002. Abstract 6820
30. Kandrup, H.E., Mahon, M.E.: Short times characterisations of stochasticity in nonintegrable galactic potentials. Astron. Astrophys. **290**, 762 (1994)
31. Kapitakinak, T.: Generating strange nonchaotic trajectories. Phys. Rev. E **47**, 1408 (1993)
32. Kaplan, J.L., Yorke, J.A.: Chaotic behaviour of multidimensional difference equations. In: Peitgen, H.O., Walter, H.O. (eds.) Functional Differential Equations and Approximations of Fixed Points. Lecture Notes in Mathematics, vol. 730, p. 204. Springer, Berlin (1979)
33. Klages, R.: Weak chaos, infinite ergodic theory, and anomalous dynamics. In: Leoncini, X., Leonetti, M. (eds.) From Hamiltonian Chaos to Complex Systems, pp. 3–42. Springer, Berlin (2013). ISBN 978-1-4614-6961-2
34. Klein, M., Baier, G.: Hierarchies of dynamical systems. In: Baier, G., Klein, M. (eds.) A Chaotic Hierarchy. World Scientific, Singapore (1991)
35. Kocarev, L., Szcepanski, J.: Finite-space Lyapunov exponents and pseudoChaos. Phys. Rev. Lett. **93**, 234101 (2004)
36. Kostelich, E.J., Kan, I., Grebogi, C., Ott, E., Yorke, J.A.: Unstable dimension variability: a source of nonhyperbolicity in chaotic systems. Physica D **109**, 81 (1997)
37. Lai, Y.C., Grebogi, C., Kurths, J.: Modeling of deterministic chaotic systems. Phys. Rev. E **59**, 2907 (1999)
38. Lepri, S., Politi, A., Torcini, A.: Chronotropic Lyapunov analysis: (I) a comprehensive characterization of 1D systems. J. Stat. Phys. **82**, 1429 (1996)
39. Lyapunov, A.M.: The General Problem of the Stability of Motion. Taylor and Francis, London (1992). English translation from the French 1907, in turn from the Russian 1892
40. Mahon, M.E., Abernathy, R.A., Bradley, B.O., Kandrup, H.E.: Transient ensemble dynamics in time-independent galactic potentials. Mon. Not. R. Astron. Soc. **275**, 443 (1995)

41. Mitchell, L., Gottwald, G.A.: On finite size Lyapunov exponents in multiscale systems. Chaos **22**, 23115 (2012)
42. Mosekilde, E.: Topics in Nonlinear Dynamics: Applications to Physics, Biology and Economic. World Scientific, Singapore (1996)
43. Moser, H.R., Meier, P.F.: The structure of a Lyapunov spectrum can be determined locally. Phys. Lett. A **263**, 167 (1999)
44. Mulansky, M., Ahnert, K., Pikovsky, A., Shepelyansky, D.L.: Strong and weak chaos in weakly nonintegrable many-body Hamiltonian systems. J. Stat. Phys. **145**, 1256 (2011)
45. Okushima, T.: New method for computing finite-time Lyapunov exponents. Phys. Rev. Lett. **91**, 25 (2003)
46. Oseledec, V.I.: A multiplicative ergodic theorem. Mosc. Math. Soc. **19**, 197 (1968)
47. Ott, E.: Chaos in Dynamical Systems. Cambridge University Press, Cambridge (1993)
48. Ott, W., Yorke, J.A.: When Lyapunov exponents fail to exist. Phys. Rev. E **78**, 056203 (2008)
49. Parisi, G., Vulpiani, A.: Scaling law for the maximal Lyapunov characteristic exponent of infinite product of random matrices. J. Phys. A **19**, L45 (1986)
50. Patsis, P.A., Efthymiopoulos, C., Contopoulos, G., Voglis, N.: Dynamical spectra of barred galaxies. Astron. Astrophys. **326**, 493 (1997)
51. Pecora, L.M., Carroll, T.L.: Synchronization in chaotic systems. Phys. Rev. Lett. **64**, 821 (1990)
52. Pesin, Y.: Dimension Theory in Dynamical Systems. Rigorous Results and Applications. Cambridge University Press, Cambridge (1997)
53. Prasad, A., Ramaswamy, R.: Characteristic distributions of finite-time Lyapunov exponents. Phys. Rev. E **60**, 2761 (1999)
54. Prasad, A., Ramaswamy, R.: Finite-time Lyapunov exponents of strange nonchaotic attractors. In: Eds. Daniel, M., Tamizhmani, K., Sahadevan, R. (eds.) Nonlinear Dynamics: Integrability and Chaos, pp. 227–234. Narosa, New Delhi (2000)
55. Ramaswamy, R.: Symmetry breaking in local Lyapunov exponents. Eur. Phys. J. B. **29**, 339 (2002)
56. Sandri, M.: Numerical calculation of Lyapunov exponents. Math. J. **6**(3), 78–84 (1996)
57. Siopis, C., Kandrup, H.E., Contopoulos, G., Dvorak, R.: Universal properties of escape in dynamical systems. Celest. Mech. Dyn. Astron. **65**, 57 (1997)
58. Smith, H., Contopoulos, G.: Spectra of stretching numbers of orbits in oscillating galaxies. Astron. Astrophys. **314**, 795 (1996)
59. Stefanski, K., Buszko, K., Piecsyk, K.: Transient chaos measurements using finite-time Lyapunov Exponents. Chaos **20**, 033117 (2010)
60. Tsiganis, K., Anastasiadis, A., Varvoglis, H.: Dimensionality differences between sticky and non-sticky chaotic trajectory segments in a 3D Hamiltonian system. Chaos, Solitons and Fractals **11**, 2281–2292 (2000)
61. Vallejo, J.C., Aguirre, J., Sanjuan, M.A.F.: Characterization of the local instability in the Hénon–Heiles Hamiltonian. Phys. Lett. A **311**, 26–38 (2003)
62. Vallejo, J.C., Viana, R., Sanjuan, M.A.F.: Local predictability and non hyperbolicity through finite Lyapunov Exponents distributions in two-degrees-of-freedom Hamiltonian systems. Phys. Rev. E **78**, 066204 (2008)
63. Vallejo, J.C., Sanjuan, M.A.F.: Predictability of orbits in coupled systems through finite-time Lyapunov exponents. New J. Phys. **15**, 113064 (2013)
64. Viana, R.L., Grebogi, C.: Unstable dimension variability and synchronization of chaotic systems. Phys. Rev. E **62**, 462 (2000)
65. Voglis, N., Contopoulos, G.: Invariant spectra of orbits in dynamical systems. J. Phys. **A27**, 4899 (1994)
66. Voglis, N., Contopoulos, G., Efthymioupoulos, C.: Method for distinguishing between ordered and chaotic orbits in four-dimensional maps. Phys. Rev. E **57**, 372 (1998)
67. Vozikis, Ch., Varvoglis, H., Tsiganis, K.: The power spectrum of geodesic divergences as an early detector of chaotic motion. Astron. Astrophys. **359**, 386 (2000)
68. Weisstein, E.W.: Lyapunov characteristic exponent, from MathWorld A Wolfram Web resource (2015). http://mathworld.wolfram.com/LyapunovCharacteristicExponent.html

69. Yanchuk, S., Kapitaniak, T.: Chaos-hyperchaos transition in coupled Rössler systems. Phys. Lett. A **290**, 139 (2001)
70. Yanchuk, S., Kapitaniak, T.: Symmetry increasing bifurcation as a predictor of chaos-hyperchaos transition in coupled systems. Phys. Rev. E **64**, 056235 (2001)
71. Yang, H.: Dependence of Hamiltonian Chaos on perturbation structure. Int. J. Bifurcation Chaos **3**, 1013 (1993)
72. Ziehmann, C., Smith, L.A., Kurths, J.: Localized Lyapunov exponents and the prediction of predictability. Phys. Lett. A **271**, 237 (2000)

# Chapter 3
# Dynamical Regimes and Time Scales

## 3.1 Temporal Evolution

There is a variety of relevant time scales when dealing with Lyapunov exponents. We have seen already the inverse of the asymptotic Lyapunov exponent, reliability time or Lyapunov time, linked to the e-folding time of the system. This quantity is a measure of the strength of chaos, and how the information of the system is preserved along a given orbit.

In addition to this time scale, there are also other time scales that provide useful insight in the studied dynamics. We have seen that the evolution of the finite-time exponents towards the asymptotic value present some oscillations when computed with very small time intervals. These are obvious variations at local scales because the ellipse axes are still subject to the local flow dynamics, without further evolution towards the most growing direction. Typically, these variations are washed out once the intervals are large enough to cover the dynamics of the flow as a whole. Good indicators of when the local regime is abandoned and we enter in the global flow are the dynamical time, $T_{\text{dyn}}$, or the Poincaré section crossing time, $T_{\text{cross}}$. Certainly, this time will depend on the selected surface of section, but it is anyway a good indicator of these time scales.

But there are variations that break any monotonically trend in the evolution of the finite-time exponents, and are sourced to long term time scales. We can cite, among others, diffusion phenomena such as Arnold diffusion, the presence of transient regular-like periods (the so-called sticky orbits) or, conversely, the presence of transient chaos. All these issues make it interesting to focus on the analysis of the distributions of the finite-time Lyapunov exponents, that are naturally well suited for tracing the variations of the flow as the intervals get longer.

---

The original version of this chapter was revised.
An erratum to this chapter can be found at DOI 10.1007/978-3-319-51893-0_5

© Springer International Publishing AG 2017
J.C. Vallejo, M.A.F. Sanjuan, *Predictability of Chaotic Dynamics*,
Springer Series in Synergetics, DOI 10.1007/978-3-319-51893-0_3

There are several studies which model universal features of the Lyapunov spectra based on the properties of an infinite set of matrices [22]. This has been done aiming to follow the evolution of the average as the time gets nearly infinite. The effective Lyapunov exponents as logarithms of products of $n$ matrices behave essentially like averages of $n$ random variables. Over short times, they are correlated, leading to a linear dependence of the cumulant generating function with $n$ [13]. Over long times, correlations may be lost. But with intermittency or in area preserving maps, there are still long time correlations, different scaling properties and multifractal structure with the sampling interval $\Delta t$. We cite among others some general results in [20, 23] or [30], some numerical findings in [28] or [31], and specifically, one Hamiltonian map scaling behaviour in [17].

Being large enough, the distribution of values will be driven uniquely by the transportation along the orbit, with no use on the linear equations of the tangent space. When dealing with effective exponents (finite but large intervals), and for hyperbolic systems, there is a simple relationship between the first and second exponents, driven by the crowding indexes. For nonhyperbolic systems, the relationship may be more complicated [13]. Usually, there is multifractality, or a strong nontrivial dependence on the order $q$ of the correlations [12].

We will focus here on the relationship between finite-time exponents when computed at "small" and "medium" intervals, where all multipliers (Lyapunov numbers) are still changing in sign and contribute to the time decay and the correlations die very slow. Instead of focusing on how the averaging evolves as the time gets longer, we will focus here on analysing the shapes of the distributions of the finite-time exponents, and how these distributions evolve when a given orbit leaves the local regime and enters into the global one.

We have described in the previous chapter of this monograph the procedure for building the finite-time Lyapunov exponents distributions. A key issue is the arbitrary initialisation of the axes once the integration is completed after the finite-time integration reaches the size $\Delta t$. The finite-time Lyapunov exponents reflect the growth rate of the orthogonal semiaxes (equivalent to the initial deviation vectors) of one ellipse centred at the initial position. These axes change their orientation and length as the orbit is integrated during a given finite-time $\Delta t$, following Eq. (2.11). When comparing these distributions, it is needed to analyse if the ordering of the exponents according to their magnitude is preserved . The definition given by Eq. (2.11) preserves the ordering as the axes evolve and a Gram-Schmidt orthonormalisation takes place along $\Delta t$. But for the shortest intervals, there is no enough time for tending to the largest growth direction, and after resetting the direction of the ellipsoid axes, the locally largest exponent may or not coincide with the previous annotated direction.

## 3.2   Regimes Identification

Each initial orientation will lead to different exponents [39]. One option is to
have the axes pointing to the local expanding/contracting directions, given by the
eigenvectors. Then, at local time scales, the eigenvalues will provide insight on the
stability of the point. Another option is to start with the axes pointing to the direction
which may have grown the most under the linearised dynamics. Yet another choice
is pointing them to the globally fastest growth direction. The technique we have
selected takes, as initial axes of the ellipse, a set of orthogonal vectors randomly
oriented, following [32]. We have made this choice because, as there is no initial
preferred orientation, the evolution of the deviation vectors is a direct consequence
of the flow time scales.

The key factor to build the finite-time distributions is finding the most adequate
$\Delta t$, to be large enough to ensure a satisfactory reduction of the fluctuations, but
small enough to reveal slow trends. This length is different for every orbit. So, in
principle, one needs to calculate the distributions for a variety of finite intervals
lengths and observe the progressive evolution of the distribution shapes. If one uses
the smallest intervals, the deviation vectors will trace the very local flow dynamics.
As one selects larger intervals, the local regime of the flow is replaced by the global
dynamics regime, and the vectors are oriented depending on the global properties of
the flow, including any transient behaviour. Finally, with the largest interval lengths,
the vectors are oriented towards the final asymptotic directions of the flow, when the
dynamics reaches the final invariant state.

In addition to the choice of the finite interval length and the initial directions
of the axes, the total integration time used to compute the distribution is also
of importance [31]. Because the integration time for gathering the finite-time
exponents is also finite, the distributions may just reflect any transient state of
the system during such integration period, instead of reflecting the global or final
stationary state. For instance, a common phenomenon found in conservative systems
is the existence of stickiness or trapped motions. A chaotic orbit may be confined to
a torus for a while, but after a very long time, it leaves the confinement and again
shows the chaotic behaviour. Depending on the total time used for gathering finite-
time exponents, different dynamics will or will not be averaged. We will expand this
discussion in the following Sect. 3.3, when dealing with transient behaviours.

We have seen in the previous chapter that when using the very smallest interval
lengths the distributions show many peaks, because the randomly oriented deviation
vectors are not able to evolve during such very small intervals. When the finite-
time intervals are slightly larger, the resulting finite-time exponent distributions
begin to be similar to flat uniform distributions. The finite-time exponents cannot
be regarded at these time scales as similar to random variables leading to Gaussian

distributions, as the deviation vectors have been allowed to evolve from the initially randomly selected deviation directions, but they had not enough time to tend to the finally fastest growing directions. These distributions are then characterised by large negative kurtosis. Finally, when the finite intervals are larger, the deviation vectors are oriented to the globally fastest growth direction, that may be or not the final asymptotic behaviour. Actually, this asymptotic direction is only reached at very long intervals.

The time scales, when the changes from the local to global behaviour are detected, can be shorter than the time scales when the asymptotic behaviour is reached and the mean of the distributions tends to the asymptotic value. This implies that when the finite-intervals are not large enough to reach the asymptotic regime, we may still detect changes due to entering in the global regime and get insight into the predictability of the orbit. This may happen even when the mean will still not be close to zero, and the fluctuations around zero will be hardly detected.

## 3.3 Transient Behaviours, Sticky Orbits and Transient Chaos

We have seen previously the existence of transient periods in the Hénon-Heiles system. This means that one orbit can behave like a regular orbit during certain periods of time, while it can show a strongly chaotic behaviour during other periods of time.

This is a behaviour similar to the intermittency phenomenon. In intermittent systems, there is an irregular alternation of periods of time between apparently regular and chaotic dynamics or between different forms of chaotic dynamics. This produces a long correlation persistence producing exponential tails in the distributions, because the trajectory behaviour follows a given pattern, and for a while, moves away from it, before returning again. This also can produce multifractal structures as the finite-time intervals are varied [6].

Another phenomenon related to the existence of transient periods of different behaviour is the existence of sticky orbits. These were firstly reported in [9]. Some points can be initially scattered in the chaotic domain but later they come close to a regular island and start moving in a regular way around it, as if they were following an invariant curve. Only after a while, their chaotic nature is again seen when they leave such a behaviour and enter in the chaotic domain again. During this period, the aperiodic orbits spend their time with a strictly zero Lyapunov exponent, if that would be exclusively computed during that time interval.

Finally, we can cite the transient chaos phenomenon, that has been widely studied by several authors. See, among others [14, 15]. The transient chaos can be found in a class of systems whose asymptotic behaviour is regular with a sensitivity to initial conditions that survives only during a finite-time interval [29]. It can also be found in open problems, linked to the existence of chaotic saddles. The nonattracting chaotic set also known as chaotic saddle or strange saddle, is formed by a set of Lebesgue measure zero of orbits that will never escape from a scattering region for both $t \to \infty$ or $t \to -\infty$ [3]. Its stable manifold contains the orbits that will never escape if $t \to \infty$, while the unstable manifold is formed by the ones that will never escape if $t \to -\infty$. The orbits that constitute the chaotic set are unstable periodic orbits, of any period, or aperiodic. Furthermore, this set is formed by the intersection of its stable and unstable manifolds, each of them being a fractal set with dimension between two and three in the three-dimensional phase space. As these two manifolds are invariant sets, also their intersection is invariant, and for that reason, all orbits that start in one point belonging to the chaotic set, will never leave the set. In fact, the stable and unstable manifolds of the chaotic set are composed of the whole set of stable and unstable manifolds of each unstable point in the chaotic set. In a region in phase space where there is a chaotic saddle, all initial conditions will escape from it after a transient, with the exception of a set of points of zero Lebesgue measure. Trajectories starting close to this set behave chaotically for a while, before diverging from it and settling into a periodic attractor [21].

The existence of these transient periods is very important when building the finite-time Lyapunov exponents, and the total integration time used to compute the distribution is the third factor affecting the distributions, in addition to the choice of the finite interval length and the initial directions of the axes as we have seen in the previous chapter [31].

Because the integration time for gathering the finite-time exponents is also finite, the distributions may just reflect any transient state of the system during such an integration period, instead of reflecting the global or final stationary state. The characterisation of the orbit may change because of the slow convergence rate towards the asymptotic global value and the finite-time characterisation of the orbit (hyperbolic or not) can change as the integration time changes [8, 19].

This means that one should take care of potentially existing transient periods within the used integration time before the final attractor is reached (dissipative system), the particle escapes (open systems) or it experiences regular-like transient periods (sticky orbits). The distributions of finite-time Lyapunov exponents can trace these periods when selecting the proper finite-time $\Delta t$ interval lengths and a total integration time constrained to the transient. But one should always take into account that this total integration time used for building the distributions must be long enough to provide enough data points for sampling and statistical analyses purposes.

## 3.4 The Hénon-Heiles System

The existence of all the above-mentioned transient periods implies that any method based on averaging certain quantities must be taken with care, and even fast convergence methods could not be fast enough to detect and characterise a given short-lived period. Averaging during long times may lead to wrong results, or at least, to ignore the existence of those transients.

There are some techniques specifically designed to cope with these cases, as for instance the AFTLE indicator, averaged finite-time Lyapunov exponents, used in [29]. These are finite-time exponents averaged on a large set of initial conditions used for estimating the duration of chaotic transients.

However, the ordinary finite-time Lyapunov exponents can trace easily this periods, with the only caveat of taking into account that the size of the finite-time interval and the total length of the integration time should produce a number of intervals enough for having a good enough distribution shape. A reduced number of intervals will lead to a distribution not useful for statistical analyses, even when some times, the analysis of the shape is enough for identification purposes.

As representative example of the above, we can follow the same approaches used in the previous chapter with the Hénon-Heiles system, and build the finite-time distributions corresponding to an orbit close to an Unstable Periodic Orbit (UPO). This orbit will initially mimic the behaviour of the periodic orbit, but, being not periodic itself, will move away from the UPO and finally escape.

This can be observed when the energy $E$ takes the value $1/4$, and set the initial conditions to be close to the Lyapunov Orbit. These initial conditions are chosen to be $x = 0.001100, y = 1.024677565117189, \dot{y} = v_y = 0.0$ (Table 3.1).

A UPO defines a frontier. Every orbit with an initial energy larger than the escape energy and moving outwards, if it crosses the Lyapunov Orbit, will escape from the system and will never come back (see [2]). The phase space of an example of such orbit is plotted in Fig. 3.1a. For the case of a UPO, each point must avoid all regions $\chi(\Delta t) < 0$. The distribution of finite-time Lyapunov exponents is formed by two peaks, both centred around positive values. When the initial condition is slightly different from the one leading to the UPO, as the selected initial condition, first row of Table 2.2, the distribution is similar to the solid line of Fig. 3.1b, where we observe two broadened peaks centred around positive values, and a tail associated with the orbit once it has escaped. When the orbit is confined, the behaviour is similar to an exact UPO and we see two peaks in the figure. The value of $\chi(\Delta t)$ oscillates between those peaks, as shown in the solid line of the smaller panel of Fig. 3.1b, leading to the intermediate spectrum of values between the main peaks. But after having integrated $8T$ time units, or after roughly 1600 finite intervals, the particle escapes. Now the range of values of $\chi(\Delta t)$ no longer oscillates, but get new values, leading to a left tail of totally different values, plotting, for instance, the smaller negative centred peak that appears at $t = 35$ time units. Indeed, as shown in

**Table 3.1** Selected orbit with transient behaviour for the Hénon-Heiles system

| Orbit | Description | Initial condition for given energy | $\lambda$ | $T_{cross}$ |
|-------|-------------|------------------------------------|-----------|-------------|
| H0 | Close to UPO | $x = 0.001100$  $y = 1.024677565117189$  $\dot{y} = v_y = 0.0$  $E = 1/4$ | – | 3.6 |

$\lambda$ is the asymptotic standard Lyapunov exponent. $T_{cross}$ is the Poincaré section crossing time corresponding to crosses with plane $x = 0$, independently of the sign of $v_x$

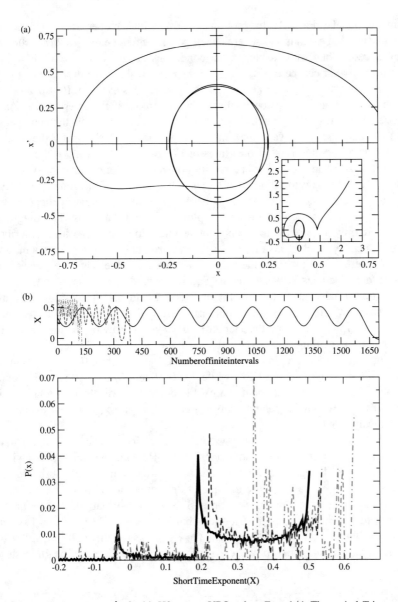

**Fig. 3.1** (**a**) Trajectory $x - \dot{x}$ of orbit H0, near a UPO, when $E = 1/4$. The period $T$ is roughly 3.6 time units. (**b**) The *solid line* shows the distribution formed with an integration of 40 time units when $\Delta t = 0.02$. The rightmost two peaks are traced when the orbit is confined, before escaping after $8T$ time units. The *dashed* probability distribution is when $\Delta t = 0.1$ and the *dotted one* when $\Delta t = 0.3$. The *smaller panel* shows the oscillating behaviour of $\chi(\Delta t)$ as the integration takes place. Taken from [31] with permission

the smaller panel of Fig. 3.1a, the motion can now follow an open track, thus the tail of the distribution extends and several small peaks centred below −0.2 (not shown) are produced. When we consider initial conditions far away from the UPO, thus orbits with smaller escape times, the general spectrum shape is different due to the tail, as it is produced by the values once the particle has escaped. But meanwhile the orbit is confined, the shape is always quite similar. If the interval size $\Delta t$ is increased, but still smaller than the escaping time, it is observed that the main peaks shift towards larger positive values and begin to merge, as shown by the dashed ($\Delta t = 0.1$) and dotted ($\Delta t = 0.3$) lines of Fig. 3.1b. As reflected in the smaller panel, the oscillation (around a larger value) of the finite time exponents values is preserved, but it begins to disappear after a smaller number of integrated intervals.

As we want to analyse further how these distributions change in shape as we vary the intervals, now will proceed to plot additional diagrams. The diagrams seen before only display the first exponent, as returned by the algorithm described in the Appendix A. But this algorithm also returns the following exponent, and it is of interest to plot how the second exponent evolves, and to check any possible dependencies between them. So, we extend the computations to the second finite-time exponent, and plot in Figs. 3.2 and 3.4 the distributions and relationship (two-dimensional distribution) $\chi_1(\Delta t) - \chi_2(\Delta t)$ diagrams for some of the prototypical orbits of the Hénon-Heiles system previously analysed. The associated numerical indexes of these distributions are found in Table 3.2.

In Fig. 3.2, we see the plots at the very local scale $\Delta t = 0.01$ for the Hénon-Heiles orbit H0, classified as a UPO, while it is confined, with no oscillation around zero. When we plot one exponent against the other one, we obtain a relationship diagram that shows a linear relationship between them, with $\chi_1$ is always expanding and $\chi_2$ always contracting.

We can also build the same plots for the close-to period-5 orbit, labeled as H4, of the Hénon-Heiles system. This is shown in Fig. 3.3. Here we see a similar behaviour. Because this orbit is regular and it does not escape, we can integrate it during longer times. Therefore, we can increase the size of the finite-interval having at the same time enough intervals for building a properly shaped distribution. So, the panels from top to bottom show the distributions built for $\Delta t = 0.01$, $\Delta t = 1.0$, and $\Delta t = 10.0$. We previously mentioned that the $T_{\text{cross}}$ was roughly 6.2 time units. We see how the distributions keep peaked below such a time scale, but begin to show

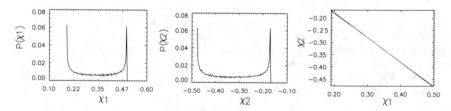

**Fig. 3.2** Finite Lyapunov exponents distributions for the Hénon-Heiles Hamiltonian and the unstable periodic orbit H0. $\Delta t = 0.01$. Adapted from [32] with permission

**Table 3.2** Numerical indexes associated with the finite Lyapunov exponent distributions corresponding to the Hénon-Heiles system, for the close-to period-5 orbit H4, for the unstable periodic orbit H0 and the weakly chaotic, cycle orbit H1, and several $\Delta t$ sizes

| $\Delta t$ | Mean | Median | $\sigma$ | $k$ | $F_+$ | $h$ |
|---|---|---|---|---|---|---|
| $\chi_1$ | | | | | | |
| UPO | | | | | | |
| 0.01 | 0.35 | 0.34 | 0.11 | −1.50 | 1.00 | 56.53 |
| Close-to period-5 | | | | | | |
| 0.01 | 0.0014 | 0.11 | 0.20 | −0.90 | 0.51 | 0.0069 |
| 1 | 0.11 | 0.11 | 0.21 | −1.22 | 0.62 | 5.13 |
| 10 | 0.077 | 0.086 | 0.047 | 1.52 | 0.92 | 70.93 |
| Weakly chaotic, cycle | | | | | | |
| 0.01 | 0.0098 | 0.051 | 0.16 | −0.70 | 0.62 | 0.78 |
| 1 | 0.074 | 0.098 | 0.17 | −1.17 | 0.64 | 5.05 |
| 10 | 0.063 | 0.072 | 0.047 | −0.31 | 0.87 | 56.37 |
| $\chi_2$ | | | | | | |
| UPO | | | | | | |
| 0.01 | −0.32 | −0.32 | 0.11 | −1.50 | 0.00 | 52.39 |
| Close-to period-5 | | | | | | |
| 0.01 | 0.0010 | −0.0097 | 0.20 | −0.90 | 0.48 | 0.0050 |
| 1 | 0.0053 | 0.028 | 0.22 | −0.95 | 0.53 | 0.22 |
| 10 | 0.022 | 0.032 | 0.066 | −0.81 | 0.61 | 10.15 |
| Weakly chaotic, cycle | | | | | | |
| 0.01 | −0.0092 | −0.049 | 0.16 | −0.70 | 0.36 | 0.74 |
| 1 | 0.00086 | 0.052 | 0.18 | −0.99 | 0.60 | 0.054 |
| 10 | 0.0042 | 0.0053 | 0.054 | −0.62 | 0.52 | 2.80 |

$\sigma$ is the standard deviation, $k$ the kurtosis, $F_+$ the probability of positivity, $h$ the hyperbolicity index

a different shape above it. Obviously, for even larger intervals, they will converge towards single-peaked distributions centred around the asymptotic values.

Finally, in Fig. 3.4, we see the plots corresponding to the weakly chaotic, cycle orbit, labeled as orbit H1. The same behaviour is seen. First, the distributions at very local scales correspond to chaotic movements. But once the intervals are larger than the $T_{\mathrm{cross}}$, the shapes change and the convergence towards the asymptotic value begins to be detected.

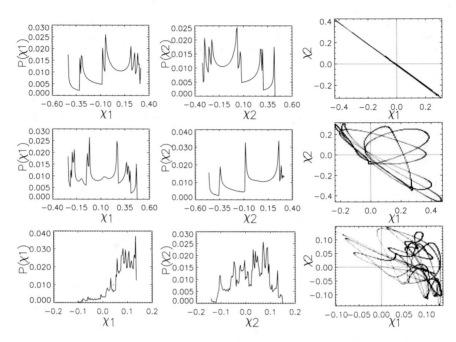

**Fig. 3.3** Finite Lyapunov exponents distributions for the Hénon-Heiles Hamiltonian and the close-to period-5, labeled as orbit H4. From *top* to *bottom*, $\Delta t = 0.01$, $\Delta t = 1$, and $\Delta t = 10$. Adapted from [32] with permission

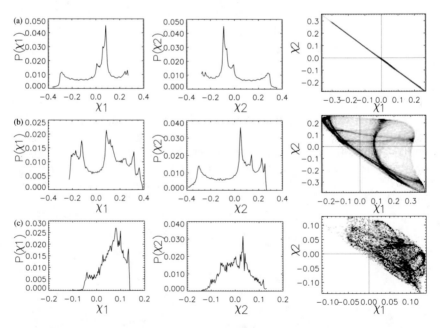

**Fig. 3.4** Finite-time Lyapunov exponents distributions for the Hénon-Heiles Hamiltonian. All panels are for the weakly chaotic, cycle orbit, labeled as H1. (**a**) $\Delta t = 0.01$, (**b**) $\Delta t = 1$, (**c**) $\Delta t = 10$. Adapted from [32] with permission

## 3.5   The Contopoulos System

Aiming to generalise these $\chi_1$–$\chi_2$ diagrams, we will plot them in an even more simpler system than the Hénon-Heiles. We will analyse another four-dimensional phase space Hamiltonian, two-degrees of freedom, originally studied by Contopoulos in [9], and given by,

$$H = \frac{1}{2}(p_x^2 + p_y^2) + \frac{1}{2}(Ax^2 + By^2) - \epsilon xy^2. \tag{3.1}$$

So, the equations of motion are the following:

$$\begin{cases} \dot{x} = p \\ \dot{y} = q \\ \dot{p} = -Ax + \epsilon xy \\ \dot{q} = -By + 2.0\epsilon xy \end{cases} \tag{3.2}$$

and the corresponding Jacobian is as follows:

$$D_v\Phi = \begin{bmatrix} 0 & 0 & 1.0 & 0 \\ 0 & 0 & 0 & 1.0 \\ -A & (2.0\epsilon y) & 0 & 0 \\ (2.0\epsilon y) & -B & 0 & 0 \end{bmatrix}. \tag{3.3}$$

This model represents two nonlinearly coupled oscillators. We have chosen it because in spite of its simplicity, it still provides a rich dynamical behaviour. In addition, it is a physically meaningful flow. The origins of this model are traced to the galactic dynamics field, like the Hénon-Heiles potential. It also belongs to the so-called galactic-type meridional potentials, reduced potentials on the meridian plane $V(R, z)$ of an axisymmetric galaxy [7].

The Contopoulos system can be seen as a simpler version of the Hénon-Heiles system, as it has only one mixed higher-order term, $xy^2$, which introduces the essential nonlinearity of the problem, $y$-axis symmetry and only two exits.

The amplitude parameters are $A = 1.6$ and $B = 0.9$. Such values are chosen to be near the resonance $\sqrt{A/B} = 4/3$ [9]. The sampled initial condition is $x = 0.03744$, $y = 0$, $\dot{x} = 0.0480$, associated with the regular motion of [10]. For this initial condition, depending on the value of the coupling parameter $\epsilon$, different orbit types are found. We have selected three values of $\epsilon$, namely, 4.4, 4.5 and 4.6. The energy value is set to $E = 0.00765$, which in the third case is close to the escape energy, given by $E_{escape} = \frac{1}{8}\frac{AB^2}{\epsilon^2}$. These orbits are listed in Table 3.3.

We focus here on comparing, for different orbits types, the distributions generated when the axes of the ellipse centred in the initial condition are allowed or not to tend to the largest stretching direction before being repointed after $\Delta t$ time units. This section describes how these distributions still provide information about the different dynamics.

**Table 3.3** Selected orbits for the Contopoulos system

| Orbit | Description | $\epsilon$ | $\lambda$ | $T_{cross}$ |
|-------|-------------|------------|-----------|-------------|
| C2 | Weakly chaotic, close to period-2, orbit | 4.5 | 0.0125 | 7.3 |
| C1 | Chaotic, between two tori, orbit | 4.4 | 0.093 | 7.0 |
| C3 | Chaotic, ergodic, orbit | 4.6 | 0.066 | 7.1 |

$\epsilon$ is the control parameter. $\lambda$ is the asymptotic standard Lyapunov exponent. The notion *weak* or *strong* chaos is associated with the relatively smaller or larger value of $\lambda$. $T_{cross}$ is the Poincaré section crossing time corresponding to crosses with plane $y = 0$, independently of the sign of $v_y$

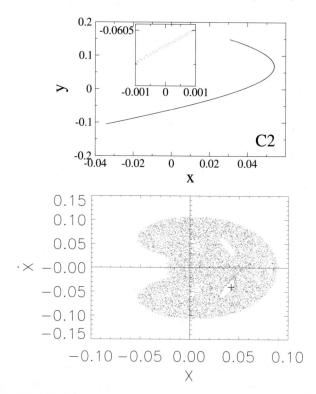

**Fig. 3.5** Orbit C2 of the Contopoulos system, a weakly chaotic orbit, that, in appearance, is a periodic orbit. It must be zoomed in to reflect its chaotic nature. The asymptotic Lyapunov exponent $\lambda = 0.0125$. *Top*: Orbit in the configuration space. *Bottom*: Poincaré section. The cross-section of another ergodic orbit ($x = 0.03$, $y = 0$, $\dot{x} = 0.04796$) has been also plotted in order to ease the visualisation of the phase portrait. Adapted from [32] and [33] with permission

The first considered orbit in this system, labeled as C2 in Table 3.3, is the case with $\epsilon = 4.5$, a weakly chaotic, close to period-2 orbit, with Poincaré section crossing-time $T_{cross} \sim 7.3$, which appears in Fig. 3.5a as a cross symbol. The density functions for the first and the second Lyapunov exponents are plotted in Fig. 3.6, and their numerical characterissssation is found in Table 3.4.

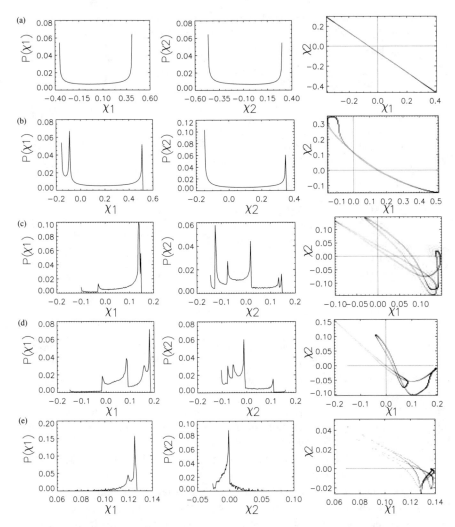

**Fig. 3.6** Finite-time Lyapunov exponents distributions for the Eq. 3.1 Hamiltonian and $\epsilon = 4.5$, weakly chaotic, close-to period-2 orbit, labeled as orbit C2, (**a**) $\Delta t = 0.01$, (**b**) $\Delta t = 1$, (**c**) $\Delta t = 7$, (**d**) $\Delta t = 10$, (**e**) $\Delta t = 100$. Adapted from [32] with permission

The distribution computed with a very short interval $\Delta t = 0.01$ appears in Fig. 3.6, panel (a). Its shows the typical shape associated with a periodic orbit. For an interval 10 times larger, we have $\Delta t = 0.1$, and the length is still below $T_{\text{cross}}$. The figures are nearly identical to the previous ones, thus they are not drawn. The fact of both being equal is not evident, as the local ellipsoid axes have now evolved a few steps, having the possibility of relaxing in the direction that permits the largest stretching, and pointing to the direction of fastest separation.

**Table 3.4** Numerical indexes associated to the finite-time Lyapunov exponent distributions corresponding to Eq. (3.1), case $\epsilon = 4.5$, close-to period-2 orbit, for several $\Delta t$ sizes

| $\Delta t$ | Mean | Median | $\sigma$ | $k$ | $F_+$ | $h$ |
|---|---|---|---|---|---|---|
| $\chi_1$ | | | | | | |
| 0.01 | 0.048 | 0.065 | 0.27 | −1.45 | 0.55 | 1.35 |
| 0.1 | 0.048 | 0.067 | 0.29 | −1.45 | 0.55 | 1.19 |
| 1 | 0.14 | 0.070 | 0.24 | −1.46 | 0.56 | 4.65 |
| 7 | 0.092 | 0.12 | 0.066 | 0.36 | 0.86 | 42.45 |
| 10 | 0.095 | 0.091 | 0.072 | 0.014 | 0.89 | 37.22 |
| 100 | 0.12 | 0.12 | 0.0084 | 11.32 | 1.00 | 3406.33 |
| $\chi_2$ | | | | | | |
| 0.01 | −0.096 | −0.11 | 0.27 | −1.45 | 0.39 | 2.69 |
| 0.1 | −0.076 | −0.095 | 0.28 | −1.44 | 0.42 | 1.87 |
| 1 | 0.082 | 0.063 | 0.18 | −1.55 | 0.56 | 4.77 |
| 7 | −0.021 | −0.028 | 0.080 | −0.74 | 0.38 | 6.66 |
| 10 | −0.02 | −0.031 | 0.057 | 0.38 | 0.22 | 12.70 |
| 100 | −0.0041 | −0.0038 | 0.011 | 4.77 | 0.19 | 68.54 |

$\sigma$ is the standard deviation, $k$ the kurtosis, $F_+$ the probability of positivity, $h$ the hyperbolicity index

For $\Delta t = 1$, panel (b), a new peak appears in the distribution of the largest exponent $\chi_1$, but the $\chi_2$ distribution remains the same. This means different rates in the evolution towards the invariant measure. When $\Delta t \sim T_{\text{cross}}$, panel (c), the $\chi_1$ distributions jump towards the positive values.

This leads to think as $T_{\text{cross}}$ as a threshold separating different regimes in the distributions, tracing local and non-local behaviour. Even when the choice of the Poincaré section is somehow arbitrary, it is based on the symmetry $y = 0$ of the potential, thus it makes sense that the crossing time for closing an orbit (if periodic) will lead to such a threshold.

At larger intervals, $\Delta t = 10$, panel (d), the oscillations around zero begin to be lost. Finally, with $\Delta t = 100$, panel (e), we are integrating several $T_{\text{cross}}$ cycles, and the distributions resemble peaks centred around the $\lambda_1 \sim 0.0125$ and $\lambda_2 \sim 0$ asymptotic Lyapunov values.

The second analysed case is a chaotic orbit, between two KAM tori, given when $\epsilon = 4.4$. This is labeled as C1 in Table 3.3. Quasi-periodic orbits are characterised by a linear divergence of neighbouring trajectories, all asymptotic exponents are zero and the motion is confined within a torus. With $\epsilon = 4.4$, the initial condition is interesting, as it does not lead to a quasi-periodic motion but to a trajectory running on a very small chaos strip between two invariant tori. The Poincaré section of this orbit appears as an elongated lobe in Fig. 3.7. The density functions for the first and the second Lyapunov exponents are plotted in Fig. 3.8. The numerical indexes which characterise such distributions are found in Table 3.5.

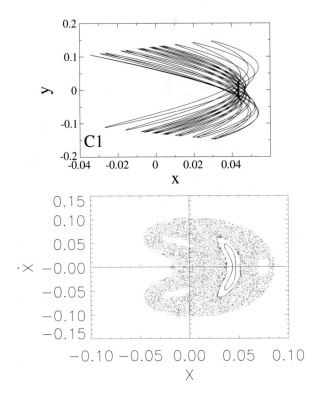

**Fig. 3.7** Orbit C1 of the Contopoulos system, a chaotic orbit with asymptotic Lyapunov exponent $\lambda = 0.093$. *Top*: Orbit in the configuration space. *Bottom*: Poincaré section. The cross-section of another ergodic orbit ($x = 0.03$, $y = 0$, $\dot{x} = 0.04796$) has been also plotted in order to ease the visualisation of the phase portrait. Adapted from [32] and [33] with permission

The main time scale to take into account seems to be again the crossing time, $T_{\text{cross}} \sim 7$. There is another physically meaningful time scale, which is the period to roughly cover the whole Poincaré section, $T_{\text{lobe}} \sim 136$.

For the shortest interval sizes, $\Delta t = 0.01$, panel (a), and $\Delta t = 0.1$, not shown, the distributions are similar, roughly double peaked, reflecting the confined motion. When the interval is increased up to $\Delta t = 1$, panel (b), there is a change in shape for $\chi_1$, with a morphology no longer similar to a periodic orbit. However, in the tangent direction, the $\chi_2$ distribution evolves at a different rate, and is still sign flipping.

Once again, as the time interval is larger than the given crossing time, for $\Delta t = 7$, panel (c), the distributions are now different. When the interval is larger than $T_{\text{cross}}$, $\Delta t = 10$, panel (d), the distributions converge to the final measure, faster for $\chi_1$. With $\Delta t = 100$, panel (e), both distributions resemble peaks centred in the asymptotic Lyapunov values $\lambda_1 \sim 0.093$ and $\lambda_2 \sim 0$.

The third analysed case is a chaotic, ergodic orbit, given when $\epsilon = 4.6$. This orbit is labeled as C3 in Table 3.3. The Poincaré section appears in Fig. 3.9. The density

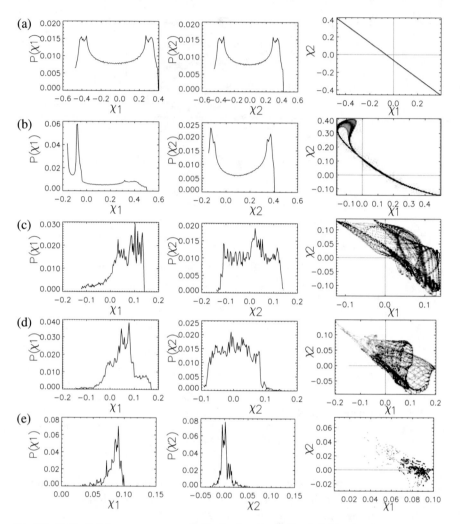

**Fig. 3.8** Finite-time Lyapunov exponents distributions for the Eq. (3.1) Hamiltonian and $\epsilon = 4.4$, chaotic orbit, between tori, labeled as orbit C1, (**a**) $\Delta t = 0.01$, (**b**) $\Delta t = 1$, (c) $\Delta t = 7$, (**d**) $\Delta t = 10$, (**e**) $\Delta t = 100$. Adapted from [32] with permission

functions for the first and the second Lyapunov exponents are plotted in Fig. 3.10. The numerical indexes which characterise such distributions are found in Table 3.6.

For the shortest intervals, $\Delta t = 0.01$, panel (a), and $\Delta t = 0.1$, not shown, both the $\chi_1$ and $\chi_2$ diagrams have widened and almost completely lost the two-peaks aspect from previous cases. For $\Delta t = 1$, panel (b), $\chi_1$ distribution changes in shape. With $\Delta t = 7$, panel (c), both distributions are almost Gaussian. This is clearly observed with $\Delta t = 10$ and $\Delta t = 100$, panels (d, e), centring around $\lambda_1 \sim 0.066$ and $\lambda_2$ values. Note, however, that even when a Gaussian shape has been achieved

**Table 3.5** Numerical indexes associated with the finite-time Lyapunov exponent distributions corresponding to Eq. (3.1), case $\epsilon = 4.4$, between tori orbit, for several $\Delta t$ sizes

| $\Delta t$ | Mean | Median | $\sigma$ | $k$ | $F_+$ | $h$ |
|---|---|---|---|---|---|---|
| $\chi_1$ | | | | | | |
| 0.01 | −0.034 | −0.033 | 0.27 | −1.42 | 0.46 | 0.92 |
| 0.1 | −0.039 | −0.038 | 0.29 | −1.42 | 0.46 | 0.93 |
| 1 | 0.077 | −0.022 | 0.20 | −1.11 | 0.47 | 3.78 |
| 7 | 0.060 | 0.070 | 0.059 | 0.34 | 0.85 | 34.88 |
| 10 | 0.054 | 0.056 | 0.052 | 0.48 | 0.84 | 40.03 |
| 100 | 0.083 | 0.086 | 0.010 | 4.51 | 0.99 | 1472.67 |
| $\chi_2$ | | | | | | |
| 0.01 | −0.013 | −0.014 | 0.27 | −1.42 | 0.48 | 0.37 |
| 0.1 | 0.012 | −0.0094 | 0.288 | −1.42 | 0.50 | 0.29 |
| 1 | 0.13 | 0.14 | 0.19 | −1.53 | 0.65 | 7.58 |
| 7 | 0.0091 | 0.014 | 0.071 | −1.07 | 0.55 | 3.66 |
| 10 | 0.0053 | 0.0049 | 0.048 | −0.91 | 0.52 | 4.54 |
| 100 | 0.0023 | 0.0010 | 0.0010 | 4.69 | 0.50 | 46.74 |

$\sigma$ is the standard deviation, $k$ the kurtosis, $F_+$ the probability of positivity, $h$ the hyperbolicity index

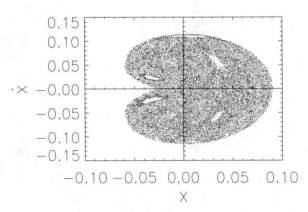

**Fig. 3.9** Poincaré section of Orbit C3 of the Contopoulos system, a chaotic orbit with asymptotic Lyapunov exponent $\lambda = 0.066$ that fullfils the available phase space. Adapted from [32] with permission

quite fast at very short intervals, the peak of $\chi_2$ is not still centred in the 0 value, implying a very low convergence of the averaging process.

This orbit is ergodic in the sense that the orbit is able to reach with the same probability all its available phase space. It is interesting to keep in mind the difference between stationarity, due to the dynamics at certain time, and ergodicity, time-averaged property of the trajectories. In a non-ergodic orbit, the trajectory does not cover the whole hypersurface of constant energy, so two different initial conditions cover different parts of the energy surface leading to different temporal averages even for times tending to infinity. In such systems there is not a unique

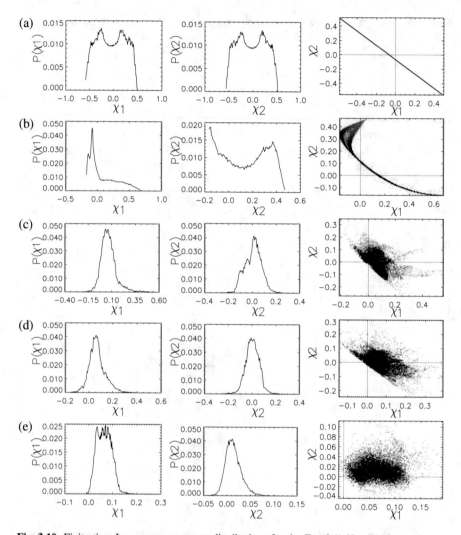

**Fig. 3.10** Finite-time Lyapunov exponents distributions for the Eq. (3.1) Hamiltonian and $\epsilon =$ 4.6, chaotic, ergodic orbit, labeled as orbit C3, (**a**) $\Delta t = 0.01$, (**b**) $\Delta t = 1$, (**c**) $\Delta t = 7$, (**d**) $\Delta t = 10$, (**e**) $\Delta t = 100$. Adapted from [32] with permission

equilibrium state, but different ones depending on the starting point. In an ergodic system a unique equilibrium state may be reached. And generic ensembles of initial conditions will evolve towards a given distribution, time-independent or with little variability on long time scales. In the case of conservative systems, there are no attractors and chaotic orbits are ergodic. But note that there may be regular-like transients in the so-called sticky orbits, where the particles wander pseudo-chaotically with strictly zero Lyapunov exponent during some time around the KAM tori. Many authors refer then to such orbits as pseudo-ergodic ones. Such transients are the reason for the broad peaks found in the distributions.

**Table 3.6** Numerical indexes associated with the finite Lyapunov exponent distributions corresponding to Eq. (3.1), case $\epsilon = 4.6$, chaotic orbit, for several $\Delta t$ sizes

| $\Delta t$ | Mean | Median | $\sigma$ | $k$ | $F_+$ | $h$ |
|---|---|---|---|---|---|---|
| $\chi_1$ | | | | | | |
| 0.01 | −0.032 | −0.031 | 0.29 | −1.18 | 0.46 | 0.74 |
| 0.1 | −0.041 | −0.037 | 0.31 | −1.14 | 0.46 | 0.84 |
| 1 | 0.091 | −0.00092 | 0.22 | −0.39 | 0.49 | 3.80 |
| 7 | 0.069 | 0.059 | 0.082 | 3.87 | 0.83 | 20.39 |
| 10 | 0.069 | 0.062 | 0.068 | 1.27 | 0.86 | 30.13 |
| 100 | 0.066 | 0.065 | 0.030 | −0.19 | 0.99 | 142.49 |
| $\chi_2$ | | | | | | |
| 0.01 | −0.015 | −0.017 | 0.29 | −1.18 | 0.48 | 0.35 |
| 0.1 | 0.014 | −0.0088 | 0.31 | −1.14 | 0.50 | 0.28 |
| 1 | 0.13 | 0.15 | 0.19 | −1.41 | 0.67 | 7.19 |
| 7 | 0.012 | 0.016 | 0.069 | 0.18 | 0.57 | 4.88 |
| 10 | 0.0067 | 0.0057 | 0.057 | 1.04 | 0.53 | 4.18 |
| 100 | 0.015 | 0.013 | 0.017 | 1.41 | 0.80 | 100.41 |

$\sigma$ is the standard deviation, $k$ the kurtosis, $F_+$ the probability of positivity, $h$ the hyperbolicity index

Now we will analyse again the relationship between the largest finite exponent, associated with the transversal direction (if allowed to evolve), and second exponent, associated with the tangential one (id), when they are calculated by re-initialising arbitrarily the axes after every interval $\Delta t$. The two-dimensional distributions histograms of the second exponent against the first one conform the third box of every row in Figs. 3.6, 3.8 and 3.10.

When comparing these distributions, it is needed to analyse if it is preserved the ordering of the exponents according to their magnitude. The definition given by Eq. (2.11) preserves the ordering as the axes evolve and a Gram-Schmidt orthonormalisation takes place along $\Delta t$. But for the shortest intervals, there is no enough time for tending to the largest growth direction, and after resetting the direction of the ellipsoid axes, the locally largest exponent may or not coincide with the previous annotated direction.

In the close to the period-2, $\epsilon = 4.5$ case, and for the smallest intervals $\Delta t = 0.01$ and $\Delta t = 0.1$, there is a linear relationship. When the local flow is expanding in one direction, it is contracting in the other one. Note a low probability region when both directions are contracting at time. For $\Delta t = 1$, the correlation is no longer linear in the $\chi_1$ contracting range. This is derived from a faster convergence rate towards the transversal direction. For the second exponent the distribution is still like a periodic one. When $\Delta t$ increases, there is a clustering of the values towards the asymptotic values.

In the $\epsilon = 4.4$ case, the results are similar for $\Delta t = 0.01$ (panel a), and $\Delta t = 0.1$ (not shown), being the density plots also linear and below the origin. When the interval is larger, $\Delta t = 1$, panel (b), we see a multivalued curve when there is expansion in the tangent direction. For $\Delta t \sim T_{\text{cross}}$, panel (c), the curve is now somehow more fuzzy. Now, the probability of finding both exponents expanding at time has increased. For $\Delta t = 10$, panel (d) the points already cluster towards the asymptotic values.

In the chaotic case, $\epsilon = 4.6$, the relationship for the smaller intervals is also linear, and when expanding, the transversal direction contracts, and vice versa. When the finite time is increased up $\Delta t \sim 1$, the relationship curve in expanding tangent direction part is more complicated. For $\Delta t = 7$ and $\Delta t = 10$ there is not correlation. For $\Delta t = 100$, the curves converge to a set of points centred in the final values.

So the relationship is linear, independently of the nature of the orbit (periodic, confined between tori or chaotic) at the very local time scales, where no evolution towards any direction is allowed. This may be a direct consequence of the arbitrary starting direction for one axis and the orthogonality of the second. But this is the same for $\Delta t = 0.1$, where many averaging steps have been performed and the vectors tend to seek the most rapidly growing directions. At these small intervals and after resetting the initial directions, the distributions still reflect the local nature of the flow, even when the finite values ordering could have been interleaved along the orbit. The comparison of the first and second distributions reflects that they essentially offer the same information.

Finally, we can mention that the linear dependence at short intervals is related to the number of degrees of freedom of the system and the associated constraints in the Lyapunov values. Indeed, in Hamiltonian systems with more degrees of freedom this linear relationship is no longer present even for the smallest intervals.

## 3.6   The Rössler System

We can see a similar evolution in the shapes of the distributions of $\chi(\Delta t)$ as the finite-time intervals $\Delta t$ grow in dissipative systems. For doing so, we will analyse again the Rössler system described in the previous chapter.

We will take as example the distributions corresponding to the point B seen in Fig. 2.7. This point is located in the hyperchaotic regime, by fixing $d = 0.25$ and $a = 0.365$

The corresponding figure showing these distributions is Fig. 3.11. It shows how the distributions shapes of the first three exponents depend on the finite interval length $\Delta t$. As $\Delta t$ increases the distributions tend to shrink, being centred around the global Lyapunov exponent. The distributions sampled a total integration time of $T = 10,000$ for all $\Delta t$, with the exception of $T = 100,000$, when $\Delta t = 100$ is analysed. The first integration time $T = 10,000$ is enough for proper display of the distributions and data analysis. Every curve contains a different but sufficient number of data points and the results are essentially the same as when using longer integration times. The case $\Delta t = 100$, however, requires the long integration $T = 100,000$, in order to have enough data points and a reliable distribution.

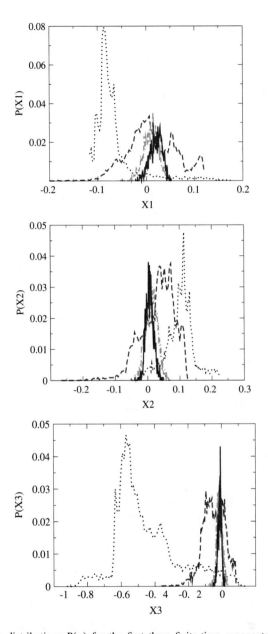

**Fig. 3.11** Density distributions $P(\chi)$ for the first three finite-time exponents corresponding to point B of Fig. 2.7, $a = 0.365$ and $d = 0.25$. These plots show how the centre and shape of the distributions depend on the finite interval length. As the finite interval $\Delta t$ is increased, the distributions tend to shrink and centre around the global Lyapunov exponent. $\Delta t = 1.0$ is *black dotted line*. $\Delta t = 10.0$ is *red dashed line*. $\Delta t = 50.0$ is *green dot-dashed line*. $\Delta t = 100.0$ is *blue continuous line*. The distributions sample a total integration time of $T = 10,000$ for all $\Delta t$, with the exception of $T = 100,000$, when $\Delta t = 100$ is analysed. Table 3.7 contains the applicable numerical indexes. Taken from [34] with permission

**Table 3.7** Numerical indexes associated with the finite-time Lyapunov exponent distributions corresponding to Fig. 3.11, for the several $\Delta t$ sizes

| $\Delta t$ | Mean | $\sigma$ | $F_+$ |
|---|---|---|---|
| $\chi_1$ | | | |
| 1.0 | −0.065 | 0.046 | 0.079 |
| 10.0 | 0.013 | 0.051 | 0.55 |
| 50.0 | 0.010 | 0.015 | 0.73 |
| 100.0 | 0.021 | 0.013 | 0.93 |
| $\chi_2$ | | | |
| 1.0 | 0.10 | 0.044 | 0.98 |
| 10.0 | 0.038 | 0.051 | 0.74 |
| 50.0 | 0.013 | 0.019 | 0.73 |
| 100.0 | 0.0071 | 0.012 | 0.71 |
| $\chi_3$ | | | |
| 1.0 | −0.44 | 0.19 | 0.028 |
| 10.0 | −0.043 | 0.065 | 0.24 |
| 50.0 | −0.013 | 0.017 | 0.21 |
| 100.0 | −0.012 | 0.011 | 0.11 |

The standard deviation is $\sigma$. The probability of positivity $F_+$

In Table 3.7 we see this trend reflected as the evolution with $\Delta t$ of the numerical indexes associated with these distributions, the very small values of $\sigma$ indicating the trend towards the asymptotic value.

The necessary time scales to make the initial axes be oriented towards the final largest growth directions can be derived from the observation of the evolution of the distributions. Actually, these time scales are different depending on the nature of the orbit. For very small time scales we have seen that there can be a linear relationship among the exponents [32]. For larger time scales, the distributions can be used to characterise the hyperbolic nature of the orbit. The tangencies among several directions can be seen as the linear dependencies between local exponents which are not lost for increasing time scales. Indeed, we could construct a local eigenvolume, that will depend on the possible dependencies among exponents, and we can follow its evolution with time. This can give us information to distinguish between chaotic and ordered orbits, as done by the GALI-k index [27], which is based on how the relationship among the different deviation vectors evolve.

## 3.7  Hyperbolicity Characterisation Through Finite-Time Exponents

A basic requirement for shadowing is hyperbolicity. Using the Rössler flow as working example, we will extend the discussion to the analysis of the possible nonhyperbolicity of a given flow and the relationship of nonhyperbolicity and the finite-time exponents distributions.

A dynamical system is hyperbolic if the phase space can be spanned locally by a fixed number of independent stable and unstable directions which are consistent under the operation of the dynamics [26] and the angle between the stable and unstable manifolds is away from zero [16, 35]. Hyperbolic systems are *structurally* stable in the sense that numerical trajectories stay close to the true ones. This phenomenon is called shadowing, as introduced earlier (Sect. 1.4).

In case of nonhyperbolicity, an orbit may not be shadowed and the computed orbit behaviour may be completely different from the true one. The nonhyperbolic behaviour can arise from tangencies between stable and unstable manifolds, from Unstable Dimension Variability (UDV), or from both.

When the nonhyperbolicity arises only from tangencies, the trajectories may be still shadowed during long times. But in a general system, we could find Unstable Periodic Orbits (UPO), KAM tori, KAM sticky orbits or chaotic sets. And in our system which is dissipative, in addition to tangencies, an attractor may pass very close to periodic orbits with different number of unstable directions. This property of unstable periodic orbits embedded in a chaotic invariant set is called Unstable Dimension Variability. In these pseudo deterministic systems, where the nonhyperbolicity arises from UDV, with or without tangencies, the shadowing may be not good, meaning that the shadowing is only valid during trajectories of a given length, sometimes very short.

The UDV indicates a variation with position of the dimension of the invariant set subspaces, and is a major difficulty when modeling high-dimensional dynamical systems because the subspaces are not invariant along a typical chaotic trajectory. The UDV can be produced by an infinite number of UPO embedded in a chaotic invariant set, having a variation with position of the dimension of the invariant set subspaces (number of eigendirections). It was first reported in the kicked double rotor [1], where the invariant set of interest is a chaotic attractor. But UDV can also appear in nonattracting chaotic sets.

Several mechanisms lead to UDV, as bubbling transition in coupled oscillators, decoherence transitions in weakly coupled or nonidentical systems, hyperchaos or extrinsic noise, with associated intermittency [5, 24, 25, 36]. Hyperchaos is a common source for UDV but note this cannot be the case in two-degrees-of-freedom (four dimensional phase space) Hamiltonians, such as the Hénon-Heiles or the Contopoulos systems. UDV seems to be common in high dimensional dynamical systems, such as coupled maps [18] and continuous flows of coupled systems [38], but it may be also present in low dimensional systems.

A sign of nonhyperbolicity and bad shadowing is then the fluctuating behaviour around zero of the finite-time exponent closest to zero [11]. This reflects in principle the varying number of dimensions along the trajectory.

So, first we need to properly identify which is the closest to zero exponent. The exponent closest to zero can be derived from the inspection of the mean $m$ of the distributions. The identification of the exponent closest to zero among all available exponents is helpful to characterise the hyperbolicity. However, we have seen how the distributions shapes change as the $\Delta t$ is varied. As a consequence, it may happen that meanwhile we are leaving the local regime, the computations will return as closest to zero a different exponent.

Aiming to explore this, we can trace a diagram indicating which is the exponent closest to zero, as returned by the algorithm we have used up to now (see Appendix A). Taking again the two coupled Rössler oscillators system as example, we can plot the index of the closest to zero exponent as derived from the mean of the probability density, for two sizes of the finite-time interval $\Delta t$. This is seen in Fig. 3.12. The index will obviously depend on the control parameters of the system, $a$ and $d$ in this case, and this is clearly seen in the figure. But what is more important, this index will depend on the size of $\Delta t$. This makes in turn to get a different set of structures in the diagrams.

For the smaller $\Delta t$ intervals, the values of the finite-time exponents have not evolved towards the final ordering. With $\Delta t = 1.0$, the directions have been already integrated 100 times, but the decorrelation has not yet taken place. For larger $\Delta t$, the distributions start to be Gaussian with a given mean centred around the global values. The top panel of Fig. 3.12 shows the index for a $\Delta t = 25$. The bottom panel corresponds to $\Delta t = 100.0$. Then, the mean of the distributions clearly tend to the global asymptotic values. As a consequence, the exponent closest to zero is the one tending to the neutral flow direction. Finally, when $\Delta t \to \infty$ the distributions will tend to be a Dirac delta function centred at the global asymptotic Lyapunov exponent value.

Once we have identified which is the closest to zero exponent, we need to detect its oscillations around zero. These oscillations can be detected when the positivity index $P_+$, or probability of getting a positive $\chi(\Delta t)$, as described by Eq. (2.12), is nearly 0.5.

How far are the positivity indexes in the parameter space $a - d$ from the 0.5 value? This proximity, or distance, is colour coded in Fig. 3.13. Here we can see that the darker regions are those with smaller values, meaning $P_+ \sim 0.5$. Conversely, the larger the values, the brighter the region and the farther from 0.5 in positive or negative directions. Areas of different behaviour of the flow, such as the upper leftmost corner, with higher coupling strengths and smaller $a$ control values, are identified even with the shorter intervals. The finest structures however can only be resolved with the larger intervals. Note that in different regions we have derived $P_+$ from different closest to zero exponents, as this identification changes along the parametric phase space, as per Fig. 3.12.

We can compare this figure with Fig. 2.7, the chart showing the hyperchaoticity of this system in the parametric space. The darker areas of Fig. 3.13, those with values of $P_+ \sim 0.5$, are the ones most likely linked to have UDV. Hyperchaos is a common source for the UDV. When comparing the darker areas with the highly chaotic areas, we see the darker zones roughly match with the hyperchaos areas.

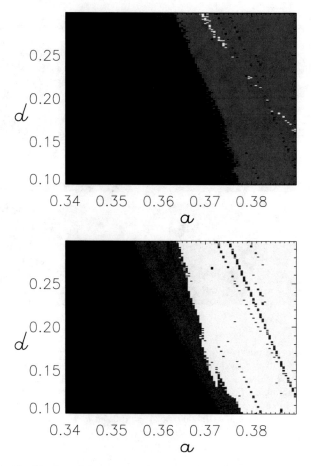

**Fig. 3.12** The identification of the exponent closest to zero varies with $\Delta t$, as derived from the mean of the probability density. *Top*: $\Delta t = 25$ and $T = 10,000$. When the exponent closest to zero is the first one we compute, it appears as *black*. If it is the second, as *red*. If the third, as *pink*. And finally for the fourth, as *white*. *Bottom*: $\Delta t = 100$ and $T = 100,000$. In this case, the exponent closest to zero is only one of the three first exponents. If it is the first, it appears as *black*. If the second, as *red*. And finally, if the third, as *white*. Taken from [34] with permission

However, the match is not perfect, and here we may conjecture that the UDV is not fully sourced to the hyperchaos here. Conversely, no area of high chaoticity matches with a high predictability area. We would like to emphasise that the exponents may fluctuate without being a clear-cut of UDV [4, 37]. There are situations where the positive tails appear not due to UDV, but rather by other mechanisms such as the quasi-tangencies between the stable and unstable manifolds near a homoclinic crisis. Nevertheless the above, the oscillations are still a good indication of the nonhyperbolic nature.

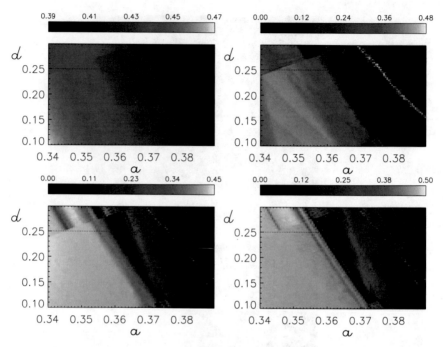

**Fig. 3.13** Probability of positivity of the closest to zero exponent, for given oscillator parameter *a* and coupling strength *d*. Scaled values give the distance to $P_+ = 0.5$. *Darker areas*, values nearly to 0.0 are those with smaller values and $P_+ \sim 0.5$. This means distributions centred around zero (stretched or shrinked). *Brighter areas* with larger values are farther from 0.5 in positive or negative direction. From *top* to *bottom*, and from *left* to *right*, $\Delta t = 1$ and $T = 10,000$, $\Delta t = 25$ and $T = 10,000$, $\Delta t = 50$ and $T = 100,000$ and $\Delta t = 100$ and $T = 100,000$. Adapted from [34] with permission

## 3.8   Concluding Remarks

We have described here how the finite-time Lyapunov exponents and their distributions serve as valid indicators on the nature of a given orbit even when the initial axes have not been pointed to any specific direction.

We have seen that the information provided by the first and second exponents, seems to be the same when computed at very local scales. At larger intervals, but below a given threshold, when axes have been allowed to point to the largest stretching direction, both exponents still trace the flow local properties, they oscillate around zero and may trace the UDV. At larger time scales, the linear relationship between both exponents is lost. We get global (averaged) values for the whole orbit or applicable domain of initial conditions, reaching the final asymptotic value at different rates.

In addition to the known fact that the more you integrate the better you can estimate the asymptotic Lyapunov exponent, we have also seen that using finite-time distributions and very short time intervals is sufficient for distinguishing the regions of different predictability behaviour.

The presence of oscillations of the closest to zero exponent is an indicator of nonhyperbolicity. This implies the necessity of the calculation of the several available exponents, as the identification of the closest one depends on the selected interval, in addition to the position in the parametric space. We have noticed for the larger intervals the exponents tend to the global values, the closest to zero points to the neutral direction and the oscillations may be then difficult of being clearly identified.

Our methods derive from calculating distributions during certain integration times T of finite exponents. This method does not use global averaged quantities during long intervals, unless strictly needed. So it can be used for open systems where transient chaos is found, taking care that the total integration time required for extracting information of the distribution is smaller than the trapping time.

We have also seen that by analysing how the shapes of the distributions change, we can detect the finite-time interval lengths when the change from the local to the global regime occurs. The Poincaré crossing time with the surface of section is a good estimator of these time scales, but unless the orbit is periodic, this crossing time depends on the selection of the surface of section. Indeed, it is not constant in the phase space once the surface has been selected.

Up to now, we have qualitatively described the changes in shapes. In the next chapter we will use a quantitative indicator, the kurtosis values of the finite-time distributions, as valid indicator. These values evolve from zero to positive values, as a consequence of the shape changes when the finite-time exponents leave the local flow dynamics and tend towards the global regime. The larger the positive kurtosis values, the more peaked the distributions will be.

Finally, one observes the asymptotic regime of the flow at the time scales when the mean of the distributions begins to be centred around the final asymptotic value [31]. As mentioned before, the flow may experience several transient periods before reaching this final asymptotic state.

# References

1. Abraham, R., Smale, S.: Non-genericity of $\Omega$-stability. Proc. Symp. Pure Math. **14**, 5 (1970)
2. Aguirre, J., Vallejo, J.C., Sanjuán, M.A.F.: Wada basins and chaotic invariant sets in the Hénon-Heiles system. Phys. Rev. E **64**, 66208 (2001)
3. Alligood, K.T., Sauer, T.D., Yorke, J.A.: Chaos. An Introduction to Dynamical Systems, p. 383. Springer, New York (1996)
4. Alligood, K.T., Sander, E., Yorke, J.A.: Three-dimensional crisis: crossing bifurcations and unstable dimension variability. Phys. Rev. Lett. **96**, 244103 (2006)
5. Barreto, E., So, P.: Mechanisms for the development of unstable dimension variability and the breakdown of shadowing in coupled chaotic systems. Phys. Rev. Lett. **85**, 2490 (2000)

6. Benzi, R., Parisi, G., Vulpiani, A.: Characterisation of intermittency in chaotic systems. J. Phys. A **18**, 2157 (1985)
7. Binney, J., Tremaine, S.: Galactic Dynamics. Princenton University Press, Princenton (1987)
8. Branicki, M., Wiggings, S.: Finite-time Lagrangian transport analysis: stable and unstable manifolds of hyperbolic trajectories and finite-time exponents. Nonlinear Proc. Geophys. **17**, 1–36 (2010)
9. Contopoulos, G.: Orbits in highly perturbed dynamical systems. I. Periodic orbits. Astron. J. **75**, 96 (1970)
10. Contopoulos, G., Grousousakou, E., Voglis, N.: Invariant spectra in Hamiltonian systems. Astron. Astrophys. **304**, 374 (1995)
11. Davidchack, R.L., Lai, Y.C.: Characterization of transition to chaos with multiple positive Lyapunov exponents by unstable periodic orbits. Phys. Lett. A **270**, 308 (2000)
12. Grassberger, P.: Generalizations of the Hausdorff dimension of fractal measures. Phys. Lett. A **107**, 101 (1985)
13. Grassberger, P., Badii, R., Politi, A.: Scaling laws for invariant measures on hyperbolic and non-hyperbolic attractors. J. Stat. Phys. **51**, 135 (1988)
14. Grebogi, C., Ott, E., Yorke, J.A.: Crises, sudden changes in chaotic attractors, and transient chaos. Physica D **7**, 181 (1983)
15. Jacobs, J., Ott, E., Hunt, R.: Scaling of the durations of chaotic transients in windows of attracting periodicity. Phys. Rev. E **56**, 6508 (1997)
16. Kantz, H., Grebogi, C., Prasad, A., Lai, Y.C., Sinde, E.: Unexpected robustness-against-noise of a class of nonhyperbolic chaotic attractors. Phys. Rev. E **65**, 026209 (2002)
17. Kottos, T., Politi, A., Izrailev, F.M., Ruffo, S.: Scaling properties of Lyapunov Spectra for the band random matrix model. Phys. Rev. E **53**, 6 (1996)
18. Lai, Y.C., Grebogi, C., Kurths, J.: Modeling of deterministic chaotic systems. Phys. Rev. E **59**, 2907 (1999)
19. Mancho, A.M., Wiggins, S., Curbelo, J., Mendoza, C.: Lagrangian descriptors: a method for revealing phase space structures of general time dependent dynamical systems. Commun. Nonlinear Sci. **18**, 3530 (2013)
20. Meiss, J.D.: Transient measures for the standard map. Physica D **74**, 254 (1994)
21. Oyarzabal, R.S., Szezech, J.D., Batista, A.M., de Souza, S.L.T., Caldas, I.L., Viana, R.L., Sanjuán, M.A.F.: Transient chaotic transport in dissipative drift motion. Phys. Lett. A **380**, 1621 (2016)
22. Parisi, G., Vulpiani, A.: Scaling law for the maximal Lyapunov characteristic exponent of infinite product of random matrices. J. Phys. A **19**, L45 (1986)
23. Prasad, A., Ramaswany, R.: Characteristic distributions of finite-time Lyapunov exponents. Phys. Rev. E **60**, 2761 (1999)
24. Sauer, T.: Shadowing breakdown and large errors in dynamical simulations of physical systems. Phys. Rev. E **65**, 036220 (2002)
25. Sauer, T.: Chaotic itinerancy based on attractors of one-dimensional maps. Chaos **13**, 947 (2003)
26. Sauer, T., Grebogi, C., Yorke, J.A.: How long do numerical chaotic solutions remain valid? Phys. Lett. A **79**, 59 (1997)
27. Skokos, Ch., Bountis, T.C., Antonopoulos, Ch.: Geometrical properties of local dynamics in Hamiltonian systems: the Generalized Alignment Index (GALI) method. Physica D **231**, 30 (2007)
28. Smith, L.A., Spiegel, E.A.: Strange accumulators. In: Buchler, J.R., Eichhorn, H. (eds.) Chaotic Phenomena in Astrophysics. New York Academy of Sciences, New York (1987)
29. Stefanski, K., Buszko, K., Piecsyk, K.: Transient chaos measurements using finite-time Lyapunov exponents. Chaos **20**, 033117 (2010)
30. Szezech Jr., J.D., Lopes, S.R., Viana, R.L.: Finite time Lyapunov spectrum for chaotic orbits of non integrable Hamiltonian systems. Phys. Lett. A **335**, 394 (2005)
31. Vallejo, J.C., Aguirre, J., Sanjuan, M.A.F.: Characterization of the local instability in the Hénon–Heiles Hamiltonian. Phys. Lett. A **311**, 26–38 (2003)

32. Vallejo, J.C., Viana, R.L., Sanjuan, M.A.F.: Local predictability and nonhyperbolicity through finite Lyapunov exponent distributions in two-degrees-of-freedom Hamiltonian systems. Phys. Rev. E **78**, 066204 (2008)
33. Vallejo, J.C., Sanjuan, M.A.F.: The forecast of predictability for computed orbits in galactic models. Mon. Not. R. Astron. Soc. **447**, 3797 (2015)
34. Vallejo, J.C., Sanjuan, M.A.F.: Predictability of orbits in coupled systems through finite-time Lyapunov exponents. New J. Phys. **15**, 113064 (2013)
35. Viana, R.L., Grebogi, C.: Unstable dimension variability and synchronization of chaotic systems. Phys. Rev. E **62**, 462 (2000)
36. Viana, R.L., Pinto, S.E., Barbosa, J.R., Grebogi, C.: Pseudo-deterministic chaotic systems. Int. J. Bifurcation Chaos Appl. Sci. Eng. **11**, 1 (2003)
37. Viana, R.L., Barbosa, J.R., Grebogi, C., Batista, C.M.: Simulating a chaotic process. Braz. J. Phys. **35**, 1 (2005)
38. Yanchuk, S., Kapitaniak, T.: Symmetry increasing bifurcation as a predictor of chaos-hyperchaos transition in coupled systems. Phys. Rev. E **64**, 056235 (2001)
39. Ziehmann, C., Smith, L.A., Kurths, J.: Localized Lyapunov exponents and the prediction of predictability. Phys. Lett. A **271**, 237 (2000)

# Chapter 4
# Predictability

## 4.1 Numerical Predictability

As introduced in the first chapter of this book, Sect. 1.4, the predictability of a system indicates how long a computed orbit is close to an actual orbit, and this concept is related to, but independent of, its stability or its chaotic nature. A system is said to be chaotic when it exhibits strong sensitivity to the initial conditions. This means that the exact solution and a numerical solution starting very close to it may diverge exponentially one from each other. The predictability aims to characterise if this numerically computed orbit may be sometimes sufficiently close to another true solution, so it may be still reflecting real properties of the model, leading to correct predictions. The real orbit is called a shadow, and the noisy solution can be considered an experimental observation of one exact trajectory. The distance to the shadow is then an observational error, and within this error, the observed dynamics can be considered reliable [27].

The shadowing property characterises the validity of long computer simulations, and how they may be *globally* sensitive to small errors. The shadowing time $\tau$ measures how long a numerical trajectory remains valid by staying close to a true orbit. The shadowing distance is the local phase space distance between both of them, as it was presented in Sect. 1.4. The shadows can exist, but it may happen that, after a while, they may go far away from the true orbit. Consequently, a proper estimation of the shadowing times is a key issue in any simulation and provides an indication about its predictability.

The shadowing time is directly linked to the hyperbolic or nonhyperbolic nature of the orbits. Hyperbolic systems are *structurally* stable in the sense that the shadowing is present during long times and numerical trajectories stay close to the

---

The original version of this chapter was revised.
An erratum to this chapter can be found at DOI 10.1007/978-3-319-51893-0_5

© Springer International Publishing AG 2017
J.C. Vallejo, M.A.F. Sanjuan, *Predictability of Chaotic Dynamics*,
Springer Series in Synergetics, DOI 10.1007/978-3-319-51893-0_4

true ones. In case of nonhyperbolicity, an orbit may be shadowed, but only for a very short time, and the computed orbit behaviour may be completely different from the true one after this period.

Non-hyperbolic behaviour can arise from tangencies (homoclinic tangencies) between stable and unstable manifolds, from unstable dimension variability or from both. In the case of tangencies, there is a higher, but still moderate obstacle to shadowing. In the so-called pseudo-deterministic systems, the shadowing is only valid during trajectories of reasonable length due to the Unstable Dimension Variability (UDV).

The UDV is reflected and quantified by the fluctuations around zero of the finite-time exponent closest to zero [11, 37]. Fluctuations around zero of the maximum transient exponent were described for attractors of quasiperiodically forced systems in [16]. Note that there are situations where the positive tails appear not due to UDV but rather by other mechanisms, such as the quasi-tangencies between the stable and unstable manifolds near a homoclinic crisis point, for example.

A key issue when there are strong obstacles to shadowing and this is only valid during short time periods is the calculation of the shadowing time $\tau$ as valid limit for the predictability of the system. This is specially relevant in high-dimensional systems, where it is hard to develop a good understanding of model accuracy or error growth.

When the shadowing times are very short, averaged quantities as Lyapunov exponents or even faster chaoticity indicators, may be handled with care, because the computed trajectories may move away from any real trajectory before the averaging time gets completed, so leading to unreliable results. In addition, we have seen in the previous Sect. 3.3, that the trajectories may also suffer transient behaviours. All these issues point to use finite exponents, able to capture the hyperbolic or not nature of the flow and to use them for estimating the shadowing times.

Once these shadowing times could have been computed, one may use them as proper averaging times. This can be applied to any averaged indicator and is of special interest in MonteCarlo simulations, based on averaging results from many initial conditions [26].

The probability distributions for the shadowing can be justified from statistical properties of the finite-time exponents [36]. The shadowing time distributions with UDV present a scaling law algebraic for small shadowing times, and exponential for large ones (longer shadowing times are exponentially improbable). The shadowing distance typically increases exponentially when change in the unstable dimension occurs [12]. Then, it decreases exponentially in the hyperbolic regions, with a lower bound determined by the computer round-off. These switches occur randomly in time, so they mimic a (biased) random walk behaviour, hence we can only give confidence to results where the amount of transversely attracting and repelling contributions nearly counterbalances (mean closest to zero) and expansions and contractions are well approximated by such a stochastic process.

The shadowing can also be described as a diffusion equation visualised as the interaction between holes, as escape routes along a given trajectory [3]. The effective range of the interactions is associated with the largest Lyapunov exponent. The shadowing is large when the holes are located in an unstable periodic orbit. The effects of the kicks in the pseudo-trajectories are included as a reflecting barrier. Such diffusion process has an equilibrium distribution leading to a shadowing time $\tau$ given by

$$\tau \sim \delta^{-h}, \tag{4.1}$$

with $\delta$ the round-off precision of the computer, and where the $h$ index is

$$h = \frac{2\|m\|}{\sigma^2}. \tag{4.2}$$

The exponent $h$ is called hyperbolicity index or predictability index, and it depends on $m$ and $\sigma$, the mean and the standard deviation of the Lyapunov exponent closest to zero. We will use this index as an indicator of the predictability of the orbits. The lowest predictability occurs when $h$ is very small and there is no improvement in $\tau$, even for large values of $\delta$. Conversely, the larger the $h$ index, the better the shadowing.

The scaling laws for $h$ are derived from the first and second cumulants [13, 28]. The variance is inversely proportional to the interval in ergodic orbits [17], algebraic powers are found when intermittency is present [25] or correlations decaying more slowly than the inverse of the time interval [30].

These scaling law is closely related to intermittency, and can be considered "intermittency in miniature". The exponential distribution is the result of small excursions that periodically move the computed trajectory away from the true trajectory, and then return towards it. The assumption is that the motion follows a biased random walk, with a drift toward a reflecting barrier. The flow sometimes goes in one direction, far away from the true solution, and sometimes moves towards it. The reflecting barrier is caused by the single-step error $\delta$, since new errors are created at each step, so the computed trajectory can never be expected to be closer than $\delta$ to the true trajectory.

## 4.2 The Predictability Index

A sign of bad shadowing is the fluctuating behaviour around zero of the closest to zero finite-time Lyapunov exponent. Plotting the finite-time distribution and assuming both the mean $m$ and the standard deviation $\sigma$ to be very small, the shadowing time $\tau$ is given by Eq. (4.2), where the exponent $h$ is the hyperbolicity index, or predictability index. We will use it as an indicator of the predictability of the orbits, with the lowest predictability linked to $h$ being very small, and a good shadowing linked to a larger $h$ index.

One important issue is to perform the $h$ computation using a closest to zero exponent, since this exponent reflects in principle the varying number of unstable dimensions along the trajectory.

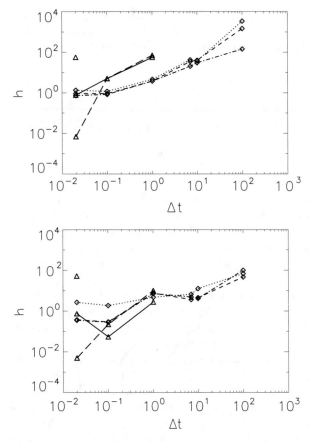

**Fig. 4.1** Hyperbolicity index, as calculated from the distributions of finite-time Lyapunov exponents, corresponding to $\chi_1$ (*upper diagram*) and $\chi_2$ (*lower diagram*). Contopoulos system values are marked by *diamonds*. *Dotted*: close-to period-2, *Dashed*: chaotic, between tori orbit, *Dashed dot*: chaotic. Hénon-Heiles system values are marked with *triangles*. *Isolated point*: UPO, *Long dashed*: close-to period-5, *Solid*: weakly chaotic, cycle. Taken from [33] with permission

But, as we have seen in the previous sections, the distributions of the finite-time Lyapunov exponents, used for deriving $h$, evolve as the $\Delta t$ intervals lengths grow. We have also seen that the distributions evolution typically changes as the lengths leave the local regime and enter in the global regime.

Aiming to check if these variations are detected by the predictability index $h$, we can calculate this index for a variety of interval lengths in both the Hénon-Heiles system, described in Sect. 2.8 and the Contopoulos system, described in Sect. 3.5.

We have computed the distributions corresponding to the first and second indexes, for a variety of finite interval lengths. The results are plotted in Fig. 4.1.

In general, $h$ grows with the interval length. For the shortest intervals, there are no Gaussian distributions and the values cannot be regarded as random variables. The exponents oscillate and $h$ keeps small, as the variance is small. For the non-UDV orbit of Henon-Heiles system, $m$ is far from zero and $h$ is large. When $h$ computed from $\chi_1$ is compared with that from $\chi_2$, the results are different even when both $\chi_1$ and $\chi_2$ fluctuate and are well correlated. The biased random walker model might not be fully applicable, but as the values are accumulated along a given orbit, they provide useful information in all orbit types, when computed from the second exponent $h(\chi_2)$.

For the largest intervals, distribution shapes are Gaussian-like, the correlations die out, and the ergodic theorem might be applied. The $h(\chi_1)$ value has a wider span of values depending on the orbit type with the larger intervals, but $h(\chi_2)$ has not.

This plot makes one to conclude that at local time scales, there are many oscillations in $h$, but as the intervals grow, it seems that is possible to differentiate between the different orbit-types and the corresponding $h$-values. For doing so, we need to make a quantitative analysis of the change of the shapes of the distributions of the finite-time Lyapunov exponents.

We aim to detect the finite-time interval lengths when the change from the local to the global regime occurs. The Poincaré crossing time with the surface of section is a good estimator of these time scales, but unless the orbit is periodic, this crossing time depends on the selection of the surface of section. Indeed, it is not constant in the phase space once the surface has been selected.

So, we will complement the $T_{cross}$ with the kurtosis values of the finite-time distributions. The kurtosis is usually considered a standard indicator of the sharpness of the peak of a frequency-distribution curve. However, it is also stated that because its definition relies on a scaled version of the fourth moment of the data, it better measures the existence of heavy tails, and not peakedness.

Because the kurtosis of the normal distribution is 3, distributions with kurtosis less than 3 are said to be platykurtic, meaning the distribution produces fewer and less extreme outliers than does the normal distribution. Distributions with kurtosis greater than 3 are said to be leptokurtic, as the Laplace distribution, for instance, that has tails that asymptotically approach zero more slowly than a Gaussian. Because of that, and for the sake of simplicity, we will actually refer as kurtosis to the actual "excess kurtosis", defined as "kurtosis$-3$". Taking into account this convention, the "kurtosis" of the normal distribution is 0.

A kurtosis above zero (i.e. excess kurtosis) is typically linked to the existence of higher peaks, because heavy-tailed distributions sometimes have higher peaks than light-tailed distributions, although this do not imply the distribution is flat-topped as sometimes reported [39].

Nevertheless the above, we are not specifically interested in detecting the existence of peaks, but indeed in detecting the departure from the local behaviour. As we have seen in the previous sections, this is typically linked to departure of a flat-top distribution merged with the creation of the tails and, possibly, the convergence towards the asymptotic value. So, the kurtosis indicator seems to be a simple and useful index for detecting these changes.

The kurtosis of the distributions of finite-time Lyapunov exponents will evolve from zero to positive values, as a consequence of the shape changes when the finite-time exponents leave the local flow dynamics and tend towards the global regime. The larger the positive kurtosis values, the more heavy-tailed and likely more peaked, the distributions will be. Finally, one observes the asymptotic regime of the flow at the time scales when the mean of the distributions begins to be centred around the final asymptotic value [32]. As mentioned before, the flow may experience several transient periods before reaching this final asymptotic state.

### 4.2.1 The Hénon-Heiles System

In this section will apply these ideas and compute the predictability index as derived from Eq. (4.2) in simple two degrees-of-freedom (d.o.f.) potentials. We will check if the subjacent diffusion model is valid in these conservative systems, where several asymptotic exponents are zero.

The Hénon-Heiles system has been detailed described in Chaps. 2 and 3. We have selected four initial conditions leading to four prototypical behaviours in this system. These orbits can be seen in Fig. 4.2, and their corresponding initial conditions are listed in Table 4.1.

The first analysed case is the orbit labeled as H1. The Poincaré section is depicted in Fig. 4.3 (top). This orbit is a weakly chaotic orbit with $\lambda = 0.015$. When considering the crosses of the $x = 0$ plane with $v_x > 0$, the averaged Poincaré section crossing time is $T_{\text{cross}} = 13.0$, with a minimum value of 10.1. When considering the crosses of the $y = 0$ plane with $v_y > 0$, the averaged Poincaré section crossing time is $T_{\text{cross}} = 16.5$, with a minimum value of 12.4. These time scales roughly indicate the change of behaviour of the finite-time distributions as the finite-time intervals grow.

In Fig. 4.3 (bottom) we have plotted the hyperbolicity index derived from the closest to zero exponent distributions and the corresponding kurtosis values, against a variety of increasing finite interval lengths $\Delta t$. The total integration time used to build the distributions is $T = 10^5$ when $\Delta t < 50.0$ and $T = 10^6$ for larger intervals sizes.

There is a clear trend of increasing $h$ values as the interval size is larger. The kurtosis shows a similar evolution from the most negative values towards the positive ones. The kurtosis curve crosses the zero value at $\Delta t = 25.1$. The corresponding finite-time Lyapunov exponents distribution of the closest to zero exponent for this interval size is seen in the inset of the figure. It is characterised by a mean $m = 0.03$ and a probability of positivity $P_+ = 0.8$. The $\Delta t$ is large enough to allow the deviation vectors to enter in the global regime of the flow, but is not large enough to reach the asymptotic zero value. Regardless of the above, some oscillations around zero are already detected and these oscillations can be considered a good indicator of the nonhyperbolicity of the flow. The predictability index derived from this distribution is $h = 54.4$. We note here that because of the

**Table 4.1** Selected orbits for the Hénon–Heiles system

| Orbit | Description | Initial condition for given energy | $\lambda$ | $< T_{cross} >$ |
|---|---|---|---|---|
| H1 | Weakly chaotic, cycle orbit | $x = 0.000000 \; y = -0.119400 \; v_x = 0.388937 \; E = 1/12$ | 0.015 | 14.8 |
| H2 | Sticky, asymptotically chaotic, orbit | $x = 0.000000 \; y = 0.095000 \; v_x = 0.396503 \; E = 1/8$ | 0.046 | 13.9 |
| H3 | Regular, close to period-1 orbit | $x = 0.000000 \; y = 0.137500 \; v_x = 0.386627 \; E = 1/12$ | 0.0 | 12.4 |
| H4 | Regular, close to period-5 orbit | $x = 0.000000 \; y = -0.031900 \; v_x = 0.307044 \; E = 1/8$ | 0.0 | 12.9 |

$\lambda$ is the asymptotic standard Lyapunov exponent. The notion *weak* or *strong* chaos is associated with the relatively smaller or larger value of $\lambda$. $< T_{cross} >$ is the averaged Poincaré section crossing time

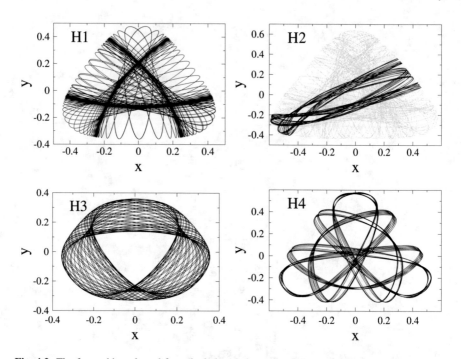

**Fig. 4.2** The four orbits selected for calculating their predictability in the Hénon-Heiles system. The corresponding initial conditions are listed in Table 4.1. *Upper left*: H1, a weakly orbit with asymptotic Lyapunov exponent $\lambda = 0.015$. *Upper right*: H2, a sticky, chaotic asymptotically, orbit, with asymptotic Lyapunov exponent $\lambda = 0.046$. The points with a regular-like transient period $t < 4000$ are plotted in darker colour. *Bottom left*: H3, a regular orbit, linked to a period 1 orbit, with asymptotic Lyapunov exponent $\lambda = 0.0$. *Bottom right*: H4, a regular orbit, linked to a period 5 orbit, with asymptotic Lyapunov exponent $\lambda = 0.0$. Taken from [35] with permission

small slope of the kurtosis and predictability curves, small changes of the estimation of the interval size does not lead to large variations in the predictability estimation.

The second analysed case is the orbit labeled as H2 in Fig. 4.2 and Table 4.1. The corresponding Poincaré section is depicted in Fig. 4.4 (top). This is a chaotic orbit with $\lambda = 0.046$. Considering the crosses of the $x = 0$ plane with $v_x > 0$, the averaged Poincaré section crossing time is $T_{\text{cross}} = 13.4$, with a minimum value of 8.9. When considering the crosses of the $y = 0$ plane with $v_y > 0$, the averaged Poincaré section crossing time is $T_{\text{cross}} = 14.5$, with a minimum value of 7.5.

In Fig. 4.4 (bottom) we observe the trend of increasing kurtosis with $\Delta t$. The kurtosis zero-cross is found at $\Delta t = 11.0$. The corresponding closest to zero exponent finite-time distribution is seen in the inset of the figure. It is characterised by a mean $m = 0.04$ and a probability of positivity $P_+ = 0.7$. The derived predictability index is $h = 20.9$. This is a worse predictability value than the previous case, yet similar in order of magnitude. We may conclude that the shadowing time scales are similar in both cases. As both orbits have positive $\lambda$ values, they are chaotic, yet predictable.

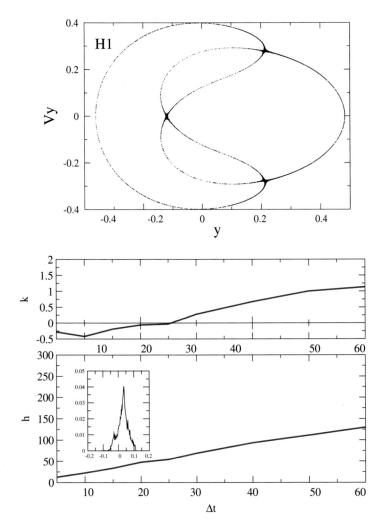

**Fig. 4.3** Hénon-Heiles weakly chaotic orbit H1. *Top*: Poincaré section $y - v_y$ with plane $x = 0$ and $v_x > 0$. *Bottom*: Evolution of the kurtosis $k$ and predictability index $h$ of the finite-time exponents distributions as the finite-time length is increased. *Inset*: Finite-time exponents distribution for $\Delta t = 25.1$. The predictability index is $h = 54.0$. Taken from [35] with permission

We have seen that orbit H1 has a relatively small Lyapunov exponent, so a relatively long Lyapunov time. This is a prototypical behaviour for a particle being chaotic, but confined to a certain region of the available phase space. But there are chaotic orbits with positive Lyapunov exponent values that show regular-like appearance during certain transient periods. These orbits stick during these transients close to islands of stability before entering in the big chaotic sea. These periods can sometimes be very short, sometimes very long. These orbits are called

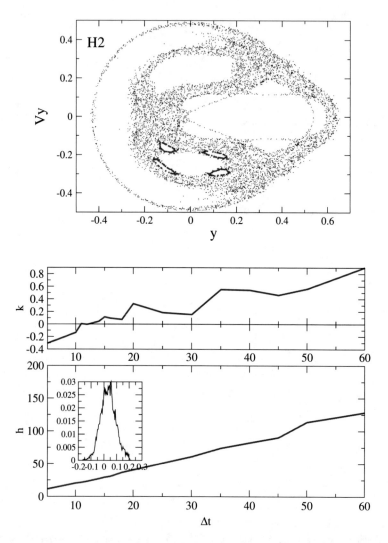

**Fig. 4.4** Hénon-Heiles sticky, chaotic asymptotically orbit H2. *Top*: Poincaré section $y - v_y$ with plane $x = 0$ and $v_x > 0$. A regular-like transient period $t < 4000$ is overplot with darker colour. *Bottom*: Evolution of the kurtosis $k$ and predictability index $h$ of the finite-time exponents distributions as the finite-time length is increased. *Inset*: Finite-time exponents distribution for $\Delta t = 11.0$. The predictability index is $h = 20.9$. Taken from [35] with permission

sticky orbits, or confined orbits [2], because they generate confined structures in the configuration space.

The sticky, chaotic asymptotically, orbit H2 presents one regular-like transient during the first 4000 time units. The Poincaré section corresponding to this period is seen in Fig. 4.5 (top). Considering the crosses of the $x = 0$ plane with $v_x > 0$, the averaged Poincaré section crossing time is $T_{\text{cross}} = 14.6$, with a minimum value of

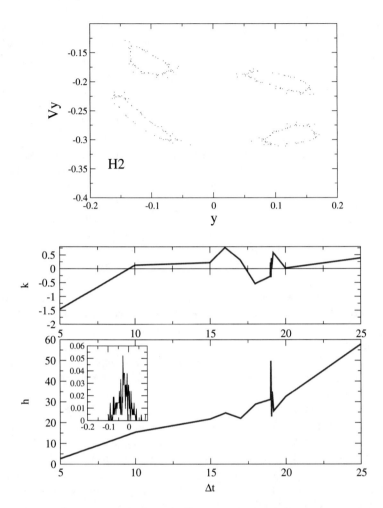

**Fig. 4.5** Regular-like period of the Hénon-Heiles chaotic orbit H2. The figure shows the points under the regular-like transient period $t < 4000$. *Top*: Poincaré section $y - v_y$ with plane $x = 0$ and $v_x > 0$. *Bottom*: Evolution of the kurtosis $k$ and predictability index $h$ of the finite-time exponents distributions as the finite-time length is increased. *Inset*: Finite-time exponents distribution for $\Delta t = 19.1$. The predictability index is $h = 31.7$. Taken from [35] with permission

13.8. When considering the crosses of the $y = 0$ plane with $v_y > 0$, the averaged Poincaré section crossing time is also $T_{\text{cross}} = 14.6$, with a minimum value of 13.2.

In Fig. 4.5 (bottom) we observe the trend of increasing kurtosis with $\Delta t$. The kurtosis zero-cross is found at $\Delta t = 19.1$. The corresponding closest to zero exponent finite-time distribution is seen in the inset of the figure. It is characterised by a mean $m = -0.01$ and a probability of positivity $P_+ = 0.28$. The derived predictability index is $h = 31.7$. This means a higher predictability during the regular-like transient when compared with the predictability value resulting from integrating beyond the transient lifetime. However, this value is lower than the value of the chaotic orbit H1. This is sourced to the selection of one of the lowest values of the available ones during the distribution shape transition, where the $h$-index values suffer several oscillations, as seen in Fig. 4.5 (bottom). But it is also sourced to the nature of the transient, that, being regular in appearance, it is not a truly regular motion.

The third analysed case is the orbit labeled as H3 in Fig. 4.2 and Table 4.1. We have chosen this orbit because we want to analyse the applicability of the power law to orbits with zero Lyapunov exponents in addition to the obvious two central trivially zero exponents. This is a regular orbit with $\lambda = 0.0$, where all exponents are zero because the Hénon-Heiles system is a 2 degrees-of-freedom Hamiltonian system.

The corresponding Poincaré section is depicted in Fig. 4.6 (top). Considering the crosses of the $x = 0$ plane with $v_x > 0$, the averaged Poincaré section crossing time is $T_{cross} = 12.4$, with a minimum value of 11.6. When considering the crosses of the $y = 0$ plane with $v_y > 0$, the averaged Poincaré section crossing time is $T_{cross} = 12.4$, with a minimum value of 11.7.

The evolution of the predictability index $h$ with the interval size is shown in Fig. 4.6 (bottom). The kurtosis shows the previous trend from the most negative values towards the positive ones. The zero crossing is found at $\Delta t = 48.3$. The corresponding distribution of the closest to zero exponent is plotted in the inset of the figure. It is characterised by a mean $m = 0.01$ and a probability of positivity $P_+ = 0.8$. The derived $h$ predictability index value is higher than the previous cases, $h = 105.5$.

When one compares the predictability of this orbit with the previous cases, the obtained predictability index $h$ is one order of magnitude larger. The biased random walk seems to be applicable to the final invariant state, even when the finite-time exponents distributions of regular orbits do not follow a normal distribution shape. This means that the test particle sometimes approaches the real orbit, having the machine precision as bias, in the contracting directions, and sometimes moves farther away from the real orbit in the expanding directions.

The fourth analysed case in the Hénon-Heiles system is the orbit labeled as H4 in Fig. 4.2 and Table 4.1. This is a regular orbit with $\lambda = 0.0$, associated with a 5th-periodic orbit. The corresponding Poincaré section is depicted in Fig. 4.7 (top). Considering the crosses of the $x = 0$ plane with $v_x > 0$, the averaged Poincaré section crossing time is $T_{cross} = 12.9$, with a minimum value of 12.3. When considering the crosses of the $y = 0$ plane with $v_y > 0$, the averaged Poincaré section crossing time is $T_{cross} = 12.9$, with a minimum value of 9.7.

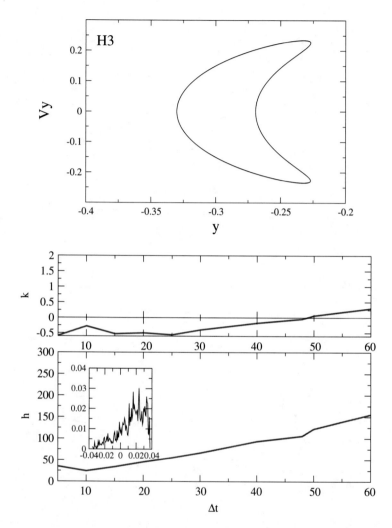

**Fig. 4.6** Hénon-Heiles regular orbit H3. *Top*: Poincaré section $y - v_y$ with plane $x = 0$ and $v_x > 0$. *Bottom*: Evolution of the kurtosis $k$ and predictability index $h$ of the finite-time exponents distributions as the finite-time length is increased. *Inset*: Finite-time exponents distribution for $\Delta t = 48.3$. The predictability index is $h = 105.5$. Taken from [35] with permission

The evolution of the predictability index $h$ with the interval size is shown in Fig. 4.7 (bottom). The previous evolution of the kurtosis from negative to positive values is observed. The kurtosis zero-cross is observed at $\Delta t = 32.1$.[1] The

---

[1] We see another zero crossing at around $\Delta t = 9.0$. This value is slightly below the $T_{\text{cross}}$ range of values. But as $\Delta t$ increases, the distribution returns to a flat shape again. As a consequence, the peaks are sourced to still be in the local regime.

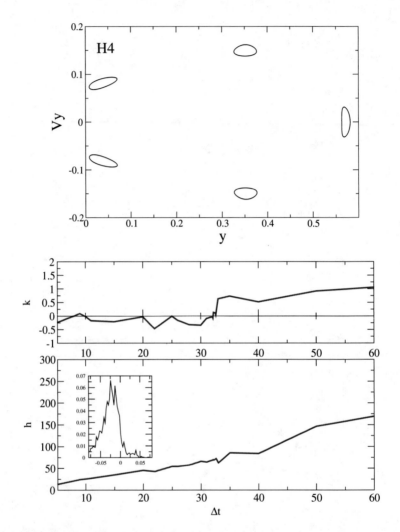

**Fig. 4.7** Hénon-Heiles regular orbit H4. *Top*: Poincaré section $y - v_y$ with plane $x = 0$ and $v_x > 0$. *Bottom*: Evolution of the kurtosis $k$ and predictability index $h$ of the finite-time exponents distributions as the finite-time length is increased. *Inset*: Finite-time exponents distribution for $\Delta t = 32.1$. The predictability index is $h = 69.2$. Taken from [35] with permission

corresponding closest to zero exponent distribution is plotted in the inset of the figure. It is characterised by a mean $m = -0.02$ and a probability of positivity $P_+ = 0.2$. Because of the Hamiltonian exponents pairing properties, the results are equivalent except for the reversed signs to the previously discussed cases.

The derived predictability index is $h = 69.2$. The predictability index is in agreement with the previous cases, with a better predictability index than the chaotic cases but a worse $h$ index than the M3, regular orbit case.

### 4.2.2 The Contopoulos System

Here, we will analyse orbits that behave in appearance like regular orbits, but that are $\lambda > 0$ chaotic. These orbits are found in the Contopoulos system [8] described in the previous chapter, Sect. 3.5. The fixed model parameters are the same, with amplitude parameters $A = 1.6$ and $B = 0.9$, as well as the initial condition $x = 0.03744$, $y = 0$, $v_x = 0.0480$, and the energy value is $E = 0.00765$.

The first case that we have considered is when we fix the control parameter $\epsilon = 4.4$. This is the orbit labeled as C1 in Table 3.3. The Poincaré section $x - v_x$ with plane $y = 0$ is seen in Fig. 4.8 (top). This is a regular in appearance, very thin chaotic strip, with $\lambda = 0.093$. When considering the crosses of the $x = 0$ plane with $v_x > 0$, the averaged Poincaré section crossing time is $T_{\text{cross}} = 13.8$, with a minimum value of 6.0. When considering the crosses of the $y = 0$ plane with $v_y > 0$, the averaged Poincaré section crossing time is $T_{\text{cross}} = 14.2$, with a minimum value of 13.9.

The evolution of the predictability index $h$ with the interval size is shown in Fig. 4.8 (bottom), where the evolution of the kurtosis from negative to positive values can be seen. The kurtosis zero-cross is found at $\Delta t = 13.9$. The corresponding closest to zero exponent distribution is plotted in the inset of the figure. It is characterised by a mean $m = -0.01$ and a probability of positivity $P_+ = 0.3$. The derived predictability index is $h = 24.2$. The predictability index is in agreement with the previous cases and this value is very similar to the predictability of the chaotic cases of the Hénon-Heiles system, confirming the lower predictability of the "regular" in appearance, chaotic orbit.

The second case analyses the same initial condition fixing $\epsilon = 4.5$. This is the orbit labeled as C2 in Fig. 4.9 and Table 3.3. The Poincaré section $x - v_x$ with plane $y = 0$ is seen in Fig. 4.9 (top). This is a weakly chaotic orbit with $\lambda = 0.0125$. This orbit is very close to a periodic orbit, meaning an averaged Poincaré section crossing time $T_{\text{cross}} = 14.5$, both for crosses of the $x = 0$ plane with $v_x > 0$, and also when considering the crosses of the $y = 0$ plane with $v_y > 0$.

The evolution of the predictability index $h$ with the interval size is shown in Fig. 4.9 (bottom). The kurtosis zero-cross in the evolution of the kurtosis from negative to positive values is found at $\Delta t = 17.15$. The corresponding closest to zero exponent distribution is plotted in the inset of the figure. It is characterised by a mean $m = -0.003$ and a probability of positivity $P_+ = 0.4$. The values of the mean and $P_+$ reflect that the asymptotic behaviour has already been reached at these time scales with contracting and expanding oscillations around zero of equal likelihood.

The figure shows strong oscillations in the predictability curve $h$ against $\Delta t$. These oscillations are linked to the presence of peaks in the distributions and the non-ergodic nature of the orbit. These oscillations make the $h$ index have strong variations with $\Delta t$, but even with these oscillations, the interval belonging to the kurtosis zero-cross is seen. The predictability index as computed from the selected $\Delta t$ is then $h = 11.9$. This predictability index is in agreement with the previous cases. This value means a lower predictability in this case than previous chaotic cases.

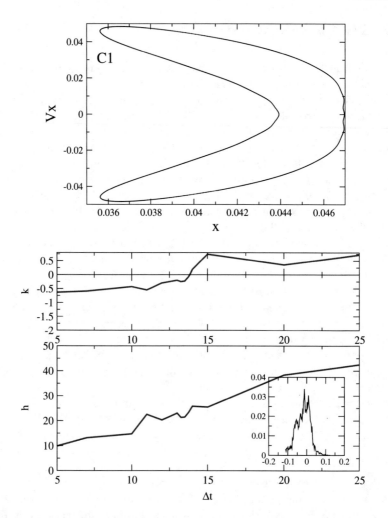

**Fig. 4.8** Contopoulos chaotic orbit C1. *Top*: Poincaré sections $x - v_x$ with plane $y = 0$ and $v_y > 0$. *Bottom*: Evolution of the kurtosis $k$ and predictability index $h$ of the finite-time exponents distributions as the finite-time length is increased. *Inset*: Finite-time exponents distribution for $\Delta t = 13.9$. The predictability index is $h = 24.2$. Taken from [35] with permission

### 4.2.3  The Rössler System

We have seen how to compute the predictability of an orbit by computing the predictability index $h$ from the finite-time distributions, following Eq. (4.2), in a variety of orbits corresponding to conservative systems. Now, we will proceed to compute the predictability in a dissipative system, the Rössler system, described in the previous chapters, Sects. 2.7 and 3.6. The initial condition is kept the same $(1, 1, 0, -1, -5, 0)$, as well as the selection of control parameters, $a$ and $d$.

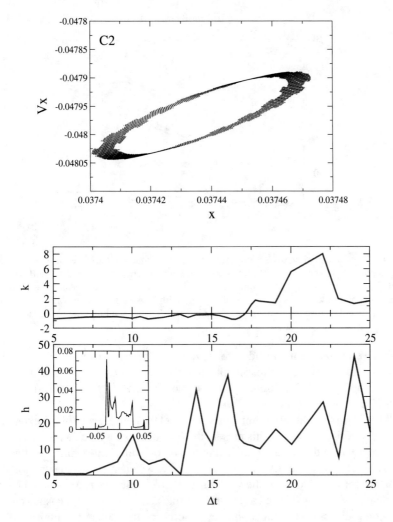

**Fig. 4.9** Contopoulos weakly chaotic orbit C2. *Top*: Poincaré sections $x - v_x$ with plane $y = 0$ and $v_y > 0$. *Bottom*: Evolution of the kurtosis $k$ and predictability index $h$ of the finite-time exponents distributions as the finite-time length is increased. *Inset*: Finite-time exponents distribution for $\Delta t = 17.15$. The predictability index is $h = 11.9$. Taken from [35] with permission

The main difference to take into account when comparing the above analysed conservative cases, and this new one is that we are dealing with orbits that will end into a chaotic attractor. In the conservative cases, every orbit may have their own time scales, and evolve towards the global regime at its own pace. In the case of an attractor, all initial conditions found in its basin of attraction will end there [1]. The time scales for all those orbits will be rather similar, depending on the nature of the attractor itself.

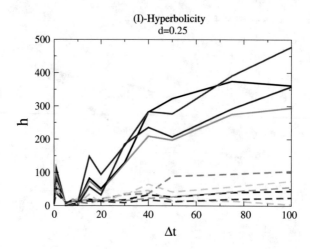

**Fig. 4.10** Hyperbolicity indexes $h$ calculated from the distributions of the closest to zero exponent, for different $\Delta t$ intervals. The control parameter $d$, coupling strength, is fixed to be $d = 0.25$. Calculations start in $a = 0.34$, every line increases $a$ in 0.05 units. *Continuous lines* are $a < 0.365$. *Dashed lines* are those with $a > 0.365$. The regimes with low and high hyperbolicity are clearly identified, but only with a large enough $\Delta t \sim 25$ interval. Taken from [34] with permission

We have computed the $h$ index for several $\Delta t$ intervals, even though Eq. (4.2) is only valid just for ergodic distributions, with a Gaussian-like shape. This can be observed in Fig. 4.10. This figure plots the evolution of $h(\Delta t)$ for a fixed value of $d = 0.25$ and several $a$ values.

When $\Delta t$ is small, the short finite times prevent the convergence of the exponents towards a limiting value. For this reason, the $h$ values do not reach a final value and consequently they do not allow to distinguish among different regimes. But as one can observe, for $\Delta t$ values larger than 25, there are two main groups of curves $h(\Delta t)$. One upper set corresponds to the values $a < 0.365$, which corresponds to the non-chaotic regime, and are plotted as continuous lines. The lower set, in dashed curves, corresponds to $a > 0.365$, containing the chaotic and hyperchaotic regimes. So, we are able to identify both regimes and it is clear that this can be achieved starting from a given time scale.

We can plot a similar diagram by fixing the control parameter $a$ and leaving free the coupling $d$. This has been done in Fig. 4.11. This figure depicts the evolution of $h$ with $\Delta t$ for fixed $a = 0.385$ and several $d$ values. For almost every coupling strength $d$, the system can be considered hyperchaotic, implying low values of $h$. However, when $d \sim 0.175$, we find only one positive exponent, implying higher values of $h$. This can be clearly observed for $\Delta t$ values larger than 40.

From observing both plots, one can conclude that there is an obvious dependency of the computed predictability not only on the combination of parameter $a$ and $d$, but also on the size of $\Delta t$, as expected. But what is more important, we have identified a threshold, common to all orbits (i.e. same initial condition, but different control

**Fig. 4.11** Hyperbolicity indexes $h$ calculated from the distributions of the closest to zero exponent, for different $\Delta t$ intervals. The control parameter $a$ is fixed to be $a = 0.385$. Calculations start in $d = 0.1$, every line increases $d$ in 0.02 units. Notice the *dashed line* $d = 0.174$, that is clearly separated from the remaining hyperchaotic cases with a large enough $\Delta t \sim 50$ interval. Taken from [34] with permission

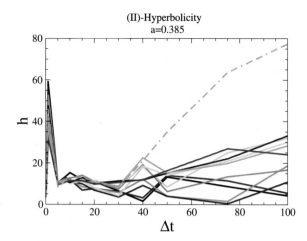

parameters), from which the predictability is different and detects the different behaviour for every orbit.

We have plotted the $h$ index against the $\Delta t$ interval lengths, for different values of the control parameters. We can reverse these plots, and we will plot now the $h$ index against the control parameters, for a variety of $\Delta t$ intervals.

This is done aiming to establish the most appropriate interval length for the computation of the hyperbolicity index. Figure 4.12 plots the evolution of $h(a)$ for different values of $\Delta t$, with a fixed value $d = 0.25$. It can be clearly observed that the hyperbolicity index $h$ decreases as $a$ increases. The black continuous curve corresponds to the larger interval size $\Delta t = 100$. The red dashed curve is $\Delta t = 50$ and the green dotted curve is $\Delta t = 25$.

The larger the time interval $\Delta t$, the higher the detail in the observed structures. This is specially relevant for detecting the lowest predictability valley at around $a = 0.36$, which is coincident with the onset of the hyperchaotic regime. We can see in the figure that one can detect the valley as from $\Delta = 25$, even when this is more clearly seen at larger intervals.

Figure 4.13 plots $h(d)$ with a fixed value $a = 0.385$, where we can see a roughly constant low predictability $h$ for any coupling strength $d$. Interestingly, even at the smallest sizes of the intervals $\Delta t$, the high predictability peak is clearly detected in this almost hyperchaotic slice.

We can conclude from both figures that for $\Delta t \sim 50$ or larger, the different regimes and thus, different predictability intervals, can be identified at sufficiently large interval lengths. Because this system is specially well suited to analyse the complex behaviour of chaotic dissipative systems, where different behaviours can be observed depending on the values of the control parameter, we extend the above results to the full parametric space $a$–$d$.

We want to see if there is a general pattern with the interval $\Delta t$, or conversely, there is no general pattern and the interplay of the control parameters mix in a complex way. Aiming to do so, we can plot the $h$ index as derived from the closest

**Fig. 4.12** Hyperbolicity
index $h$ against $a$, calculated
from the distributions of the
closest to zero exponent, for
different $\Delta t$ intervals. The
*black continuous curve* is
$\Delta t = 100$. The *red dashed* is
$\Delta t = 50$ and the *green dotted*
is $\Delta t = 25$. Fixed coupling
strength $d = 0.25$. The
general trend of $h$ decreasing
with $a$ is observed at all
intervals, but the details are
better seen with larger $\Delta t$.
Taken from [34] with
permission

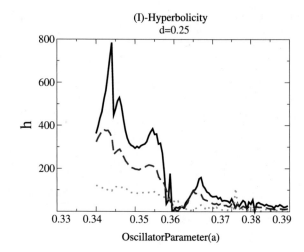

**Fig. 4.13** Hyperbolicity
index $h$ against $d$, calculated
from the distributions of the
closest to zero exponent, for
different $\Delta t$ intervals. The
*black continuous curve* is
$\Delta t = 100$. The *red dashed* is
$\Delta t = 50$ and the *green dotted*
is $\Delta t = 25$. Fixed oscillator
parameter $a = 0.385$. The
high predictability peak at
$d \sim 0.17$ is better seen with
the largest $\Delta t$. Taken from
[34] with permission

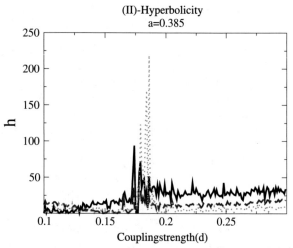

to zero exponent in the full parametric space $a$–$d$, for different $\Delta t$ values. This can
be seen in Fig. 4.14. Note that from this perspective, the plots of Figs. 4.12 and 4.13
are slices of the whole parametric space numerical explorations of Fig. 4.14.

We can observe from this figure that when using the smaller $\Delta t$, which in
principle is associated with the less reliable $h$ predictability values, there are still
regions which are identified as having different predictability behaviour. This plot
allows us to identify different predictability zones even for the smaller intervals in
certain areas of the parametric space.

When inspecting Fig. 4.14 we can also see that two main different behaviour
areas are clearly visible, as the available parametric space is divided into two
behaviour regions (left and right) from $\Delta t = 25$ onwards. Indeed, some specific
regions can be differentiated as having a different behaviour even at $\Delta t = 1$,

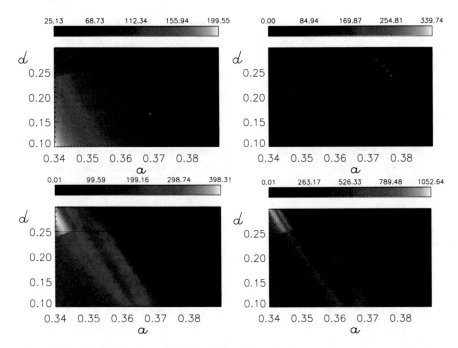

**Fig. 4.14** Predictability chart, or *h* index derived from the closest to zero exponent, for given oscillator parameter *a* and coupling strength *d*. *Darker values* reflect *h* lower and means poor predictability. From *top* to *bottom*, and from *left* to *right*, $\Delta t = 1$ and $T = 10{,}000$, $\Delta t = 25$ and $T = 10{,}000$, $\Delta t = 50$ and $T = 100{,}000$ and $\Delta t = 100$ and $T = 100{,}000$. Taken from [34] with permission

although this identification is not very clear in this extreme case. This is the case of the upper-and-leftmost corner of the *a*–*d* diagram, corresponding to the higher coupling *d* and lower *a* values, identified as a region behaving differently than the others, with a decorrelation time very short, even with the shortest intervals. Other regions are however only clearly identified at larger $\Delta t$, when the distributions are nearly Gaussian, and both *m* and $\sigma$ are small enough. This means that the decorrelation time for reaching a Gaussian-like shape and reliable *h* indexes vary with the *a*–*d* values. Some regions are easily identified as having a different predictability behaviour for shorter $\Delta t$ values than other regions, where larger $\Delta t$ are needed.

These results (predictability charts) focused on finding the interval sizes for detecting the nonhyperbolic cases of worst predictability. We can get some additional insight into the sources of the nonhyperbolicity by comparing these predictability *h* charts with the hyperchaoticity charts and positivity charts that were previously seen in Sect. 2.7, Fig. 2.7, and Sect. 3.6, Fig. 3.13, respectively.

The nonhyperbolicity can arise from tangencies between stable and unstable manifolds, from UDV or from both. When UDV is present, the shadowing times can be very short with oscillations around zero of the closest to zero exponent are present. By comparing Figs. 3.13 and 4.14, the predictability $h$ charts with the positivity charts, and because the latter reflect the around-zero oscillations of the closest to zero exponent, this comparison may provide a clue to the role of UDV in the loss of predictability. As starting point, our system is very close to the one showed in [40] and [41], where UDV was reported to be present. So UDV is likely the source for the nonhyperbolicity, at least in the cases of worst predictability (smaller shadowing times). We see that there is a good agreement among the darkest areas of both figures, mainly the central part, where both $P_+ \sim 0.5$ and $h$ is low. Now, we should be aware that at the largest intervals, the $P_+$ is not properly detected, as the distributions are tending towards the asymptotic global value. Again, $\Delta t \sim 25$ seems to be an adequate range for comparison. Some regions of different behaviour, as the one conforming the right part of the parametric space, are nevertheless detected with almost every $\Delta t$ interval. In this region, we obtain low predictability $h$ but there are no large oscillations around zero, as reflected on how $P_+$ deviates from 0.5.

Hyperchaos is a common source for UDV. When comparing the worst predictability areas, or darker areas of Fig. 4.14, with the high chaotic areas of Fig. 2.7, we see the darker zones roughly match with the hyperchaos areas of Fig. 2.7. However, the match is not perfect, and here we may conjecture the UDV is not fully sourced to hyperchaos here. Conversely, no area of high chaoticity matches with a high predictability area. When comparing the high predictability areas, or brighter areas in the rightmost column of Fig. 4.14, with the less chaotic areas of Fig. 2.7 (those with none or just one single positive exponent) we note they are similar, but not identical. This means that not all well-behaved areas have the same order of predictability. As discussed previously, these comparisons are best when using the largest $\Delta t$. But even at $\Delta t \sim 25$ or even less, the chart can be of interest.

### 4.2.4  A Galactic System

Along this chapter we have seen how one can derive the predictability index from the distributions of the finite-time Lyapunov exponents and by analysing how their shapes evolve with the interval length $\Delta t$. This has been done in simple conservative systems, such as the Hénon-Heiles and the Contopoulos system, and also in dissipative systems, such as the one formed by two non-linearly coupled Rössler oscillators.

We will show now how the computations of the predictability index behaves when applied to a more realistic, three degree-of-freedom, six-dimensional system.

When dealing with high dimensional systems, some issues must be taken into account. In systems with 2 or less degrees-of-freedom, regular and non-regular orbits are separated by impenetrable barriers, the KAM-tori, leading to islands of regularity embedded into a surrounding chaotic sea. According to the KAM theorem, these tori will survive under small perturbations if their frequencies are sufficiently incommensurable [19]. Resonant tori may be strongly deformed even under small perturbations, however, leading to a complicated phase space structure of interleaved regular and chaotic regions. Where tori persist, the motion can still be characterised in terms of $N$ local integrals. Where tori are destroyed, the motion is chaotic and the orbits move in a space of higher dimensionality than $N$.

In systems with more than 2 degrees-of-freedom, like the selected potential, the chaotic sea contains a hidden non-uniformity because the motion can diffuse through invariant tori, reaching arbitrarily far regions. Within the chaotic sea there are cantori, leaking or fractured KAM-tori, associated with the breakdown of integrability. These cantori are just partial barriers that, over short times, divide the chaotic orbits into two types: confined and non-confined. The confined ones are chaotic orbits which are trapped near the regular islands and, for a while, exhibit regular-like behaviour. Conversely, the unconfined orbits travel unmixed through the whole allowed sea. Furthermore, the cantori are partial barriers, allowing one orbit to change from one class to the other, via the intrinsic diffusion or Arnold diffusion. This is a very slow phenomenon, with typical time scales longer than the age of the Universe.

In six-dimensional phase space systems, the sticky transients are not present, cantori appears, and the Arnold diffusion produces the ultimate merging of all orbits. But this diffusion seems to be very small. Strong local instability does not mean diffusion in phase space. And some chaotic trajectories may require very long time scales to reveal its asymptotic nature. They can have very short Lyapunov times but they cannot show the expected significant orbital changes but at long times [4, 7, 31]. In these cases, the dynamics can be considered as regular motion from an astronomical point of view during the applicable time scales.

As a consequence, the fact that two regions in phase space are connected does not mean that all the areas in that volume will be accessed on comparable time scales. This long lifetime transient, unconfined orbit, is sometimes called near-invariant distribution, as it uniformly populates the filling region. It is remarkable that even when the true equilibrium corresponds to a uniform distribution through both cavities, at physically meaningful time scales, the quasi-equilibrium may have one cavity uniformly populated while the other one is, essentially, empty.

The selected model is taken again from the Astronomy field, a model of our Galaxy (i.e. the Milky-Way), based on a mean field potential. This selection could be seen as too simplistic. The gravitational $N$-body simulation is a common tool to study the evolution of the galaxies and the formation of their features. The galaxy is modelled as a self-gravitating system containing stars, gas particles and dark matter, all of them modelled as point-like masses. The self-consistency of these models captures very well the necessary details of the galactic dynamics, however, the available computational resources impose a limit to the number of particles to be taken into account. This usually implies an artificial smoothing of the potential and a proper handling of the required scaling parameters.

As an alternative, another approach that can be taken is the use of simulations based on a single mean field potential. As there are no collisions among particles, the dynamics of a galaxy can be considered to be formed by independent trajectories within the global potential where the motion of each star is just driven by a continuous smooth potential. A dynamical model usually mathematically describes the potential as a function of the distance from the centre of the galaxy. Some potentials are derived at specific snapshots of the $N$-body simulations and some others are selected to physically represent desired characteristics of the galaxies.

There is a considerable number of realistic galactic models in the literature that capture and describe several observed features in galaxies such as bars, spirals or rings. See, among others, [24, 29, 38] and [9]. In this chapter, we have selected the potential described in [18] and the references therein. This is a smooth fixed gravitational time-independent potential that models the Milky Way, but focuses on the parameters controlling the shape and orientation of a triaxial dark halo. It consists of a Miyamoto-Nagai disk [23], a Hernquist spheroid and a logarithmic halo.

This potential provides enough information to be considered to be a realistic one. It does not take the gravitational influence of a rotating galactic bar into account, but it is considered sufficient because it reproduces the flat rotation curve for a Milky Way type galaxy and can be easily shaped to the axial ratios of the ellipsoidal isopotential surfaces. By selecting different values of the model parameters, it will allow focusing on their effect on the predictability of the model. These control parameters are the orientation of the major axis of the triaxial halo and the flattening.

The dynamical system to solve is a particle (star) subject to a potential built upon three components: $V = \Phi_{\text{disk}} + \Phi_{\text{sphere}} + \Phi_{\text{halo}}$. The respective contribution of every component to the gravitational potential is given by:

$$\Phi_{\text{disk}} = -\alpha \frac{GM_{\text{disk}}}{\sqrt{R^2 + (a + \sqrt{z^2 + b^2})^2}}, \tag{4.3}$$

$$\Phi_{\text{sphere}} = -\alpha \frac{GM_{\text{sphere}}}{r + c}, \tag{4.4}$$

$$\Phi_{\text{halo}} = v_{\text{halo}}^2 \ln \left( C_1 x^2 + C_2 y^2 + C_3 xy + (z/q_z)^2 + r_{\text{halo}}^2 \right), \tag{4.5}$$

where the various constants $C_1$, $C_2$ and $C_3$ are given by:

$$C_1 = \left( \frac{\cos^2 \phi}{q_1^2} + \frac{\sin^2 \phi}{q_2^2} \right), \tag{4.6}$$

$$C_2 = \left( \frac{\cos^2 \phi}{q_2^2} + \frac{\sin^2 \phi}{q_1^2} \right), \tag{4.7}$$

$$C_3 = 2 \sin \phi \cos \phi \left( \frac{1}{q_1^2} - \frac{1}{q_2^2} \right). \tag{4.8}$$

It must be noted that there is no symmetry in the potential and $V(\phi)! = V(-\phi)$ because of the sign dependency in the $xy$ coupling factor $C3$. When $\phi = 0$, $q_1$ is aligned with the Galactic $X$-axis and Eq. (4.5) reduces to

$$\Phi_{halo} = v_{halo}^2 \ln \left( (x/q_1)^2 + (y/q_2)^2 + (z/q_z)^2 + r_{halo}^2 \right). \quad (4.9)$$

The results with $\phi = 0$ are then comparable with non-triaxial, purely logarithmic potentials. When $\phi = 90$, $q1$ is aligned with the Galactic $Y$-axis and it takes the role of $q2$. The parameter $\alpha$ could range from 0.25 up to 1.0, and following [18] and [15], is fixed to 1.0. We also adopt $M_{disk} = 1.0 \cdot 10^{11} M_{sun}$, $M_{sphere} = 3.4 \cdot 10^{10} M_{sun}$, $a = 6.5$ kpc, $b = 0.26$ kpc, $c = 0.7$ kpc, $r_{halo} = 12$ kpc. We have also fixed $v_{halo} = 128$ km/s (leading to a Local Standard of Rest LSR of 220 km/s). The time units are in Gyr with these parameter values.

The control parameters of this model are the orientation of the major axis of the triaxial halo $\phi$ and its flattening. This flattening is introduced along the three axes by the parameters $q_1$, $q_2$, $q_z$. The $q_z$ represents the flattening perpendicular to the Galactic plane, while $q_1$ and $q_2$ are free to rotate in the Galactic plane at an angle $\phi$ to a right-handed Galactocentric $X$, $Y$ coordinate system. We follow the parameter settings of [18] and, without loss of generality, $q_2 = 1.0$, $q_1 = 1.4$ and $q_z = 1.25$.

Regarding the particle initial conditions, we use stars with velocities within the halo kinematics range [5, 6]. These initial conditions, and the values of the control parameter $\phi$, corresponding to the four analysed orbits, are listed in Table 4.2. The initial velocity vector in all cases is contained into the $z = 0$ plane, meaning $v_z = 0.0$, and is normal to the $x$ axis, meaning $v_x = 0.0$. We just select in every initial condition the velocity modulus, $|v| = v_y$.

Massless particles subject to the selected gravitational potentials are integrated using the standard variational method described in the Appendix A to compute the finite-time Lyapunov exponents. We solved at the same time the flow equations and the fundamental equations or evolution of the distortion tensor, associated with the initial set of deviation vectors used for the exponents computation. The detailed equations are not listed here for the sake of conciseness, but are easily derived following the same methods described for the remaining simpler systems. We use as integrator the well-known and reliable Dop853 algorithm described in [14]. The Lyapunov exponents follow the pairing property and that the energy value is constant throughout the computation; typically having a percentual error of $10^{-8}$ for the potential.

We will analyse the four orbits shown in Fig. 4.15, labeled as M1, M2, M3 and M4. They are listed in Table 4.2. The first orbit, M1, is a regular orbit, selected for comparing the time scales of this model with the previously analysed meridional potentials. The following orbits, M2, M3 and M4, are chaotic orbits. These are confined within some phase space domain for a while, but, afterwards, can scape from those domains. As a consequence of these transients, the distribution shapes vary depending on the selection of the finite-time lengths.

The first analysed case is a regular orbit, characterised by $\lambda = 0.0$, and confined into the disk plane $z = 0$ for the whole integration. This is the orbit labeled as M1 in Fig. 4.15 and in Table 4.2. We have selected it in order to compare the predictability time scales in this model with respect to the meridional potentials seen before. This

**Table 4.2** Selected orbits for the 3 d.o.f. Milky-Way system

| Orbit | Orbit type | Initial condition | Control parameter | $\lambda$ |
|---|---|---|---|---|
| M1 | Regular | $x = 10.0 \; y = 0.0 \; z = 0.0 \; v_x = 0.0 \; v_y = 200.0 \; v_z = 0.0$ | $\phi_{\text{halo}} = 0.0$ | 0.0 |
| M2 | Chaotic | $x = 10.0 \; y = 0.0 \; z = 10.0 \; v_x = 0.0 \; v_y = 45.0 \; v_z = 0.0$ | $\phi_{\text{halo}} = 0.0$ | 0.14 |
| M3 | Chaotic | $x = 10.0 \; y = 0.0 \; z = 10.0 \; v_x = 0.0 \; v_y = 200.0 \; v_z = 0.0$ | $\phi_{\text{halo}} = 90.0$ | 0.099 |
| M4 | Strongly chaotic | $x = 5.0 \; y = 0.0 \; z = 0.5 \; v_x = 0.0 \; v_y = 100.0 \; v_z = 0.0$ | $\phi_{\text{halo}} = 0.0$ | 5.86 |

$\lambda$ is the asymptotic standard Lyapunov exponent. The notion *weak* or *strong* is associated with the relatively smaller or larger value of $\lambda$

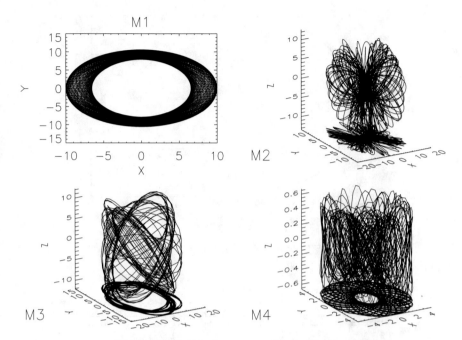

**Fig. 4.15** Four orbits selected for calculating the predictability in a Milky-Way type potential. The initial conditions and halo orientation values are listed in Table 4.2. *Upper left*: M1, a regular orbit confined to the disk, with asymptotic Lyapunov exponent $\lambda = 0.0$. *Upper right*: M2, a chaotic orbit out of the disk plane, with asymptotic Lyapunov exponent $\lambda = 0.14$. *Bottom left*: M3, a chaotic orbit out of the disk plane, with asymptotic Lyapunov exponent $\lambda = 0.099$. *Bottom right*: M4, a strongly chaotic orbit, inner and close to the disk plane, with asymptotic Lyapunov exponent $\lambda = 5.86$. Taken from [35] with permission

is of interest because a single-step error $\delta t$ may have different consequences in every model, and the shadowing times for regular orbits in different models are not necessarily similar. The corresponding Poincaré section $y - v_y$ with plane $x = 0$ is seen in Fig. 4.16 (top). When considering the crosses of the $x = 0$ plane with $v_x > 0$, the averaged Poincaré section crossing time is $T_{cross} = 0.50$, with a minimum value of 0.44. When considering the crosses of the $y = 0$ plane with $v_y > 0$, the averaged Poincaré section crossing time is $T_{cross} = 0.50$, with a minimum value of 0.46.

The evolution of the kurtosis and predictability index with the interval size is shown in Fig. 4.16 (bottom). Conversely to the regular orbits seen in the meridional potential cases, the kurtosis does not show a simple trend as the interval length grows, and there is a set of different zero crossings starting around $\Delta t = 0.06$. These oscillations at small intervals lengths below the $T_{cross}$ range of values are sourced to the fluctuations of the shapes of the distributions when the intervals are very small [33]. The $T_{cross}$ indicates when the global regime is reached, and the kurtosis zero-cross corresponding to these scales is seen at $\Delta t = 0.6$. The corresponding closest to zero exponent distribution is plotted in the inset of Fig. 4.16. It is characterised by a mean $m = 6.53$ and a probability of positivity $P_+ = 0.99$. The mean and the

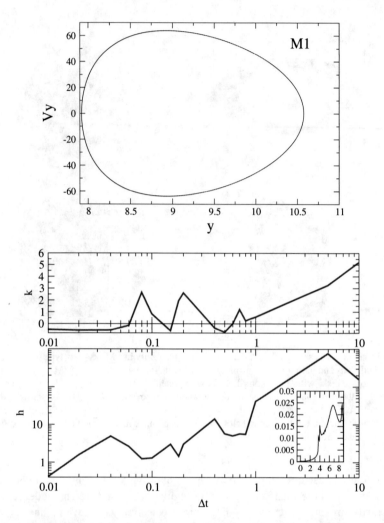

**Fig. 4.16** Milky Way regular orbit M1, confined into the disk plane, with asymptotic Lyapunov exponent $\lambda = 0.0$. *Top*: Poincaré sections $y - v_y$ with plane $x = 0$ and $v_x > 0$. *Bottom*: Evolution of the kurtosis $k$ and predictability index $h$ of the finite-time exponents distributions as the finite-time length is increased. *Inset*: Finite-time exponents distribution for $\Delta t = 0.6$. The predictability index is $h = 5.03$. Taken from [35] with permission

probability of positivity indicate that we have detected the global regime, but we are still far away from the asymptotic regime.

The predictability index is $h = 5.03$. Note that this is a very low predictability when one compares it with the values seen in the meridional potentials, both for regular and chaotic orbits. This indicates one must handle with care long integrations in this potential. The shadowing time of a regular orbit can be large or small depending on the analysed potential, because of the different dynamical times. And

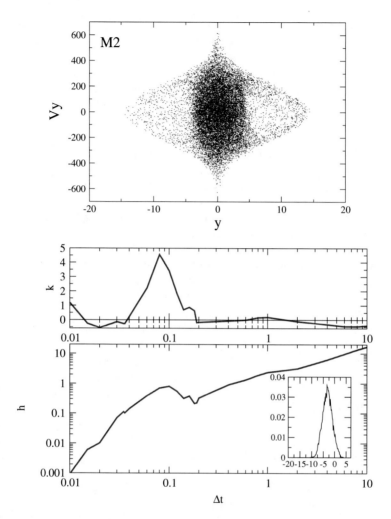

**Fig. 4.17** Milky Way chaotic orbit M2, out of the disk plane, with asymptotic Lyapunov exponent $\lambda = 0.14$. *Left*: Poincaré sections $y - v_y$ with plane $x = 0$ and $v_x > 0$. *Right*: Evolution of the kurtosis $k$ and predictability index $h$ of the finite-time exponents distributions as the finite-time length is increased. *Inset*: Finite-time exponents distribution for $\Delta t = 0.6$. The predictability index is $h = 1.31$. Taken from [35] with permission

when the shadowing times are very low, one should use higher precision schemes, even when the gain in shadowing time may be small in the extreme cases.

The next analysed initial condition corresponds to the star labeled as M2 in Fig. 4.15 and Table 4.2. The Poincaré section $y - v_y$ with plane $x = 0$ corresponding to this orbit is seen in Fig. 4.17 (top). This is a chaotic orbit characterised by $\lambda = 0.14$. We have selected this orbit because it is a chaotic orbit that initially remains in a limited domain of the phase space, but then fills up a larger domain of the available phase space, as seen in Fig. 4.17 (top) . When considering the crosses of the $x = 0$ plane with $v_x > 0$, the averaged Poincaré section crossing time is $T_{\text{cross}} = 0.61$, with a minimum value of 0.32. When considering the crosses of the $y = 0$ plane with $v_y > 0$, the averaged Poincaré section crossing time is $T_{\text{cross}} = 0.53$, with a minimum value of 0.34. When considering the crosses of the $z = 0$ plane with $v_z > 0$, the averaged Poincaré section crossing time is $T_{\text{cross}} = 0.41$, with a minimum value of 0.04.

The evolution of the kurtosis and predictability index $h$ with the interval size is shown in Fig. 4.17 (bottom). Conversely to previous models, there is not a simple increasing trend of kurtosis with $\Delta t$. Instead, there is a set of different zero crossings. We observe a zero crossing in the kurtosis curve at around $\Delta t = 0.035$, but this value is well below the $T_{\text{cross}}$ range of values. There is also a zero crossing at a very large interval size (not shown in the figure), when the asymptotic regime is reached. The kurtosis zero-cross corresponding to the time scales when the global regime of the flow is reached is seen at $\Delta t = 0.6$. The corresponding closest to zero exponent distribution is plotted in the inset of the figure. It is characterised by a mean $m = -2.8$ and a probability of positivity $P_+ = 0.08$. The mean and probability of positivity indicate that we have detected the global regime, but we are still very far away from the asymptotic regime.

The predictability index is $h = 1.31$. Note that this is a very low predictability when compared with previous cases, indicating that some care must be taken when performing long integrations using this potential. Indeed, taking into account the kurtosis oscillations, we may consider that we have taken an upper limit for the value of the predictability, and within certain transients, the predictability of the orbit may be even worse.

The following initial condition is the orbit labeled as M3 in Fig. 4.15 and Table 4.2. The Poincaré section of this orbit is seen in Fig. 4.18 (top). This orbit is characterised by $\lambda = 0.099$. The movement is then chaotic, with some transient periods spent in the external lobes of the section. When considering the crosses of the $x = 0$ plane with $v_x > 0$, the averaged Poincaré section crossing time is $T_{\text{cross}} = 0.87$, with a minimum value of 0.71. When considering the crosses of the $y = 0$ plane with $v_y > 0$, the averaged Poincaré section crossing time is $T_{\text{cross}} = 0.90$, with a minimum value of 0.76. When considering the crosses of the $z = 0$ plane with $v_z > 0$, the averaged Poincaré section crossing time is $T_{\text{cross}} = 0.86$, with a minimum value of 0.62.

The evolution of the predictability index $h$ with the interval size is shown in Fig. 4.18 (bottom). The zero crossing of the kurtosis within the range of values indicated by the Poincaré crossing time $T_{\text{cross}}$ is found at $\Delta t = 1.01$. The corresponding finite-time distribution is plotted in the inset of the figure. It is characterised by a mean $m = 0.83$ and a probability of positivity $P_+ = 0.8$. The derived predictability index is $h = 2.06$. Similar to the previous case, we can consider this value as an upper limit to the predictability of the orbit, since the orbit may suffer transient periods with an even worse predictability.

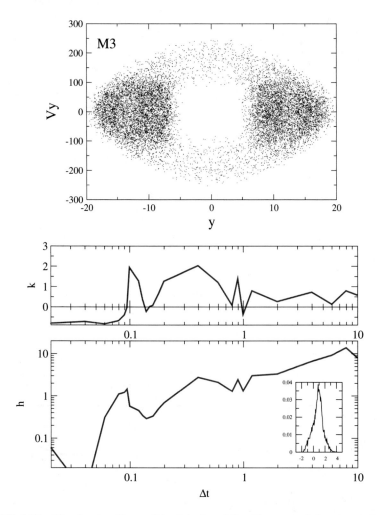

**Fig. 4.18** Milky Way chaotic orbit out of the disk plane, M3, with asymptotic Lyapunov exponent $\lambda = 0.099$. *Top*: Poincaré sections $y - v_y$ with plane $x = 0$ and $v_x > 0$. *Bottom*: Evolution of the kurtosis $k$ and predictability index $h$ of the finite-time exponents distributions as the finite-time length is increased. *Inset*: Finite-time exponents distribution for $\Delta t = 1.01$. The predictability index is $h = 2.06$. Taken from [35] with permission

The fourth analysed condition is the orbit labeled as M4 in Fig. 4.15 and Table 4.2. This initial condition corresponds to a star located close to the disk plane, in an inner region than the previous orbits. The Poincaré section of this orbit is seen in Fig. 4.19 (top). This orbit is characterised by $\lambda = 5.86$. The movement is then strongly chaotic. When considering the crosses of the $x = 0$ plane with $v_x > 0$, the averaged Poincaré section crossing time is $T_{\text{cross}} = 0.18$, with a minimum value of 0.12. When considering the crosses of the $y = 0$ plane with $v_y > 0$, the averaged

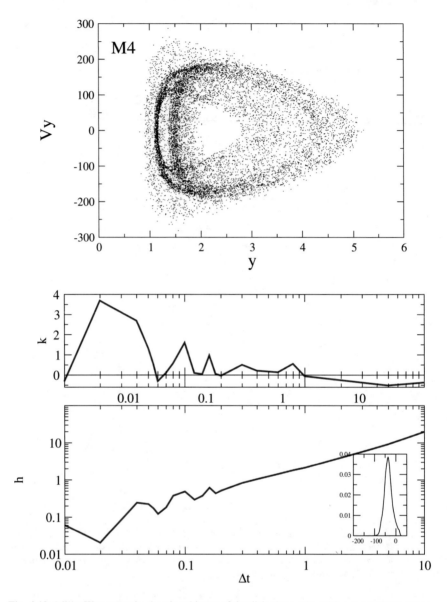

**Fig. 4.19** Milky Way strongly chaotic orbit out of the disk plane, M4, with asymptotic Lyapunov exponent $\lambda = 5.86$. *Top*: Poincaré sections $y - v_y$ with plane $x = 0$ and $v_x > 0$. *Bottom*: Evolution of the kurtosis $k$ and predictability index $h$ of the finite-time exponents distributions as the finite-time length is increased. *Inset*: Finite-time exponents distribution for $\Delta t = 0.07$. The predictability index is $h = 0.18$. Taken from [35] with permission

Poincaré section crossing time is $T_{cross} = 0.19$, with a minimum value of 0.16. When considering the crosses of the $z = 0$ plane with $v_z > 0$, the averaged Poincaré section crossing time is $T_{cross} = 0.095$, with a minimum value of 0.06.

The evolution of the predictability index $h$ with the interval size is shown in Fig. 4.19 (bottom). The zero crossing of the kurtosis within the range of values indicated by the Poincaré crossing time $T_{cross}$ is found at $\Delta t = 0.07$. The corresponding finite-time distribution is plotted in the inset of the figure. It is characterised by a mean $m = -33.55$ and a probability of positivity $P_+ = 0.043$. We are again far away from the time scales when the asymptotic dynamics is reached. The derived predictability index is $h = 0.18$. This is a very low value when compared with the previous cases, in agreement with the relatively high Lyapunov asymptotic exponent.

## 4.3   Concluding Remarks

This chapter deals with the forecast of predictability, and not with the forecast of chaoticity. Both terms are closely related, but they do not always follow the same trend. We have estimated the predictability index for a variety of prototypical orbits in several conservative galactic potentials. Instead of focusing on the *reliability* time as the inverse of the asymptotic Lyapunov exponent, thus in the chaotic, or not, nature of the orbits, we analyse the *predictability* of the system, understood as a measure of its shadowing properties.

We have seen how analysing the changes in the shapes of the distributions one can derive the predictability index. The finite-time Lyapunov exponents distributions reflect the underlying dynamics [32], and by using arbitrarily oriented deviation axes, one can detect varying the finite-time interval lengths, when there is a change from the local to the global, not yet asymptotic regime. The key issue here is that the finite-time Lyapunov exponents are computed with an initial random orientation of the ellipse axes, that is reset once the finite interval is integrated. This allows to obtain different predictability indexes that will detect the cross from local to global regime [33].

A sign of bad shadowing is the fluctuating behaviour of the closest to zero of the available Lyapunov exponents. In a general case, there can be several exponents tending to zero. Following the methods presented in [34] for a dissipative system, one should increase the finite-time interval length and select the closest to zero for deriving the predictability index. In dissipative systems the finite interval size where there is a change from local to global is the same, because all close enough orbits end in the same attractor, evolving towards similar time scales. But in conservative systems, there are no attractors, and the finite-time lengths are specific to every orbit. We have calculated these lengths in conservative systems by computing the Poincaré crossing times and detecting changes in the sign of the kurtosis of the finite-time distributions.

The presence of the oscillations of the closest to zero exponent is an indicator of nonhyperbolicity. This implies the necessity of the calculation of several available exponents, as the identification of the closest one depends on the selected interval, in addition to the position in the parametric space. We have noticed that for the larger intervals the exponents tend to the global values, the closest to the zero points to the neutral direction and the oscillations may be then difficult of being clearly identified.

The results presented here are of general interest in describing how the predictability index computed using Eq. (4.2) provides information on the system dynamics. When calculating predictability indexes, one must take into account the time scales of the analysed system for better interpretation of the range of values corresponding to a given model. Regarding regular orbits, we see in this table that the shadowing times can be very different when comparing regular orbits belonging to different models. This is because the consequences of a single-step error $\delta t$ are different depending on the model. Regarding chaotic orbits, we have seen that the predictability indexes of chaotic orbits can be also different when they belong to different models. Two orbits can be chaotic, yet one may have a larger index than the other. The predictability index is linked to the hyperbolic nature of the orbit, and in turn, to its energy and stiffness of the system. The existence of two or more time scales in different directions, one quickly growing, one slowly growing, can lead to stiffness, and the finite-time exponents reflect these expanding/contracting behaviours. In addition, the predictability indexes depend on the time scales when there is a change on these behaviours, and the global regime is reached. Different energy values lead to different dynamical times, so to different time scales.

Finite-time Lyapunov exponents techniques are indeed useful for studying those transient periods that the dynamics may suffer before ending in a final invariant state. The distributions can be built using shorter total integration times than those required for reaching the asymptotic behaviour. There is a limitation, however, when reducing the total integration time, that is the number of finite intervals needed for having good statistics values derived from the distribution. As $\Delta t$ increases, the number of intervals needed for building a well-sampled distribution and a reliable mean, deviation or kurtosis calculation, also increases.

Regarding the selection of the finite-time interval sizes, we have observed that, conversely to the dissipative case, the time scales when the deviation vectors leave the local regime of the flow and begin to evolve under the global dynamics can be different and smaller than the time scales where the asymptotic regime starts. When $\Delta t$ is large enough, the distributions tend to shrink and centre around the asymptotic value. In early works of [10], the interval length (characteristic time) for the effective exponents was $t_H$. We have seen that it is possible to use intervals smaller than $t_H$ for gaining insight into the properties of the flow.

The method we have used indicates the most adequate interval length for estimating the predictability of a given orbit, independent of their regular or irregular nature. When there are several dynamical transients, reflected in changes in the shapes of the finite-time distributions, and consequently, different zero-crossings in the kurtosis curves, this method returns upper limits to the orbit predictability.

The dynamical times are different depending on the studied orbit and model. The predictability indexes values can be used for comparing the predictability of different orbits. They reflect how the shadowing time increases as the precision in the computations increases. Low predictability indexes lead to short shadowing times. Selecting an integration scheme and assuring the energy is kept constant in time (within some small error) does not imply the calculated orbit is shadowed by a real one beyond certain limits.

The predictability index estimates this shadowing time duration. A given numerical scheme with certain precision can be enough when the shadowing times are large. But this may be not the case when the shadowing times are shorter. A high predictability index may indicate that high-precision time-consuming schemes are not necessary, even for chaotic orbits. Indeed, RK4 integrators provide good results even for the strongest chaotic orbits seen in the presented meridional potentials. Conversely, a low predictability index points to the use of more powerful schemes, as required in the Milky Way model, for instance. In a general case, when these indexes are really small, large increases in precision does not mean large increases in shadowing times, and one should consider the cost of implementing more complex and time-consuming schemes.

The percentages of regular and chaotic orbits in the phase space is not only a function of its spatial location but also a function of the total energy and main parameters of the model [21], and the amount of chaotic and regular motions in a given ensemble of initial conditions is related to the forecast of its predictability. Chaos detection methods based on saturation or averaging return different values as the saturation times vary because of the possible evolving presence of different regions of chaos, moderate or strong [20]. We have estimated the finite-time lengths to use in the calculation of the $h$-index, from the analysis of changes in the shapes of the distributions. When several zero-crossings are present, we have selected the zero based on the Poincaré section crossing time scales

Our method has been applied to regular and chaotic orbits, and the results point to the validity of the shadowing times returned by Eq. (4.2) even when the ergodic diffusion model may not be fully applicable in the regular cases. Our work has focused on the predictability index as estimator of the accuracy of an orbit in some time-independent potentials. The time independence of such potentials allows the trajectories to be either periodic, regular or chaotic (strong or weak). But the only unsual transitions found are those when a chaotic trajectory behaves like a regular orbit and requires long time scales to reveal its true chaotic nature. In time-dependant potentials, one can find migrations from chaotic to regular [22]. The presented techniques are applicable both when there are changes from regular to chaotic motions, or changes from chaotic to regular motions, as in the time-dependant cases. This is because the predictability index presented in this paper derives from solving the variational equations and detecting changes in the shapes of the finite-time Lyapunov exponents as the finite-time intervals are increased.

# References

1. Alligood, K.T., Sauer, T.D., Yorke, J.A.: Chaos. An Introduction to Dynamical Systems, p. 383. Springer, New York (1996)
2. Athanassoula, E., Romero-Gómez, M., Bosma, A., Masdemont, J.J.: Rings and spirals in barred galaxies - III. Further comparisons and links to observations. Mon. Not. R. Astron. Soc. **407**, 1433 (2010)
3. Buljan, H., Paar, V.: Many-hole interactions and the average lifetimes of chaotic transients that precede controlled periodic motion. Phys. Rev. E **63**, 066205 (2001)
4. Cachucho, F., Cincotta, P.M., Ferraz-Mello, S.: Chirikov diffusion in the asteroidal three-body resonance (5, -2, -2). Celest. Mech. Dyn. Astron. **108**, 35 (2010)
5. Casertano, S., Ratnatunga, K.U., Bahcalli, J.N.: Kinematic modeling of the galaxy. II - Two samples of high proper motion stars. Astrophys. J. **357**, 435 (1990)
6. Chiba, M., Beers, T.C.: Structure of the galactic stellar halo prior to disk formation. Astrophys. J. **549**, 325 (2001)
7. Cincotta, P.M., Giordano, C.M.: Topics on diffusion in phase space of multidimensional Hamiltonian systems. In: New Nonlinear Phenomena Research, p. 319. Nova Science, Hauppauge (2008)
8. Contopoulos, G.: Orbits in highly perturbed dynamical systems. I. Periodic orbits. Astron. J. **75**, 96 (1970)
9. Contopoulos, G., Harsoula, M.: 3D chaotic diffusion in barred spiral galaxies. Mon. Not. R. Astron. Soc. **436**, 1201 (2013)
10. Contopoulos, G., Grousousakou, E., Voglis, N.: Invariant spectra in Hamiltonian systems. Astron. Astrophys. **304**, 374 (1995)
11. Davidchack, R.L., Lai, Y.C.: Characterization of transition to chaos with multiple positive Lyapunov exponents by unstable periodic orbits. Phys. Lett. A **270**, 308 (2000)
12. Do, Y., Lai, Y.C.: Statistics of shadowing time in nonhyperbolic chaotic systems with unstable dimension variability. Phys. Rev. E **69**, 16213 (2004)
13. Grassberger, P., Badii, R., Politi, A.: Scaling laws for invariant measures on hyperbolic and non-hyperbolic attractors. J. Stat. Phys. **51**, 135 (1988)
14. Hairer, E., Norsett, S.P., Wanner, G.: Solving Ordinary Differential Equations, I, Nonstiff Problems, 2nd edn. Springer, Berlin (1993)
15. Johnston, K.V., Spergel, D.N., Hernquist, L.: The disruption of the sagittarius dwarf galaxy. Astrophys. J. **451**, 598 (1995)
16. Kapitaniak, T.: Distribution of transient Lyapunov exponents of quasiperiodically forced systems. Prog. Theor. Phys. **93**, 831 (1995)
17. Kottos, T., Politi, A., Izrailev, F.M., Ruffo, S.: Scaling properties of Lyapunov spectra for the band random matrix model. Phys. Rev. E **53**, 6 (1996)
18. Law, D.R., Majewski, S.R., Johnston, K.V.: Evidence for a triaxial Milky Way dark matter halo from the sagittarius stellar tidal stream. Astrophys. J. Lett. **703**, L67 (2009)
19. Lichtenberg, A.J., Lieberman, M.A.: Regular and Chaotic Dynamics. Applied Mathematical Sciences, vol. 38, 2nd edn. Springer, New York (1992)
20. Maffione, N.P., Darriba, L.A., Cincotta, P.M., Giordano, C.M.: Chaos detection tools: application to a self-consistent triaxial model. Mon. Not. R. Astron. Soc. **429**, 2700 (2013)
21. Manos, T., Athanassoula, E.: Regular and chaotic orbits in barred galaxies - I. Applying the SALI/GALI method to explore their distribution in several models. Mon. Not. R. Astron. Soc. **415**, 629 (2011)
22. Manos, T., Machado, R.E.G.: Chaos and dynamical trends in barred galaxies: bridging the gap between N-body simulations and time-dependent analytical models. Mon. Not. R. Astron. Soc. **438**, 2201 (2014)
23. Miyamoto, M., Nagai, R.: Three dimensional models for the distribution of mass in galaxies. Publ. Astron. Soc. Jpn. **27**, 533 (1975)
24. Pfenniger, D.: The 3D dynamics of barred galaxies. Astron. Astrophys. **134**, 373 (1984)

25. Prasad, A., Ramaswany, R.: Characteristic distributions of finite-time Lyapunov exponents. Phys. Rev. E **60**, 2761 (1999)
26. Sauer, T.: Shadowing breakdown and large errors in dynamical simulations of physical systems. Phys. Rev. E. **65**, 036220 (2002)
27. Sauer, T., Grebogi, C., Yorke, J.A.: How long do numerical chaotic solutions remain valid? Phys. Lett. A **79**, 59 (1997)
28. Sepulveda, M.A., Badii, R., Pollak, E.: Spectral analysis of conservative dynamical systems. Phys. Lett. **63**, 1226 (1989)
29. Skokos, Ch., Patsis, P.A., Athanassoula, E.: Orbital dynamics of three-dimensional bars - I. The backbone of three-dimensional bars. A fiducial case. Mon. Not. R. Astron. Soc. **333**, 847 (2002)
30. Tomsovic, S., Lakshminarayan, A.: Fluctuations of finite-time stability exponents in the standard map and the detection of small islands. Phys. Rev. E **76**, 036207 (2007)
31. Tsiganis, K., Varvoglis, H., Hadjidemetriou, J.D.: Stable chaos in high-order Jovian resonances. Icarus, **155**, 454 (2002)
32. Vallejo, J.C., Aguirre, J., Sanjuan, M.A.F.: Characterization of the local instability in the Henon-Heiles Hamiltonian. Phys. Lett. A **311**, 26 (2003)
33. Vallejo, J.C., Viana, R.L., Sanjuan, M.A.F.: Local predictability and nonhyperbolicity through finite Lyapunov exponent distributions in two-degrees-of-freedom Hamiltonian systems. Phys. Rev. E **78**, 066204 (2008)
34. Vallejo, J.C., Sanjuan, M.A.F.: Predictability of orbits in coupled systems through finite-time Lyapunov exponents. N. J. Phys. **15**, 113064 (2013)
35. Vallejo, J.C., Sanjuan, M.A.F.: The forecast of predictability for computed orbits in galactic models. Mon. Not. R. Astron. Soc. **447**, 3797 (2015)
36. Viana, R.L., Pinto, S.E., Barbosa, J.R., Grebogi, C.: Pseudo-deterministic chaotic systems. Int. J. Bifurcation Chaos Appl. Sci. Eng. **11**, 1 (2003)
37. Viana, R.L., Barbosa, J.R., Grebogi, C., Batista, C.M.: Simulating a chaotic process. Braz. J. Phys. **35**, 1 (2005)
38. Wang, Y., Zhao, H., Mao, S., Rich, R.M.: A new model for the Milky Way bar. Mon. Not. R. Astron. Soc. **427**, 1429 (2012)
39. Westfall, P.H.: Kurtosis as peakedness, 1905–2014, R.I.P. Am. Stat. **68**, 191 (2014)
40. Yanchuk, S., Kapitaniak, T.: Symmetry increasing bifurcation as a predictor of chaos-hyperchaos transition in coupled systems. Phys. Rev. E **64**, 056235 (2001)
41. Yanchuk, S., Kapitaniak, T.: Chaos-hyperchaos transition in coupled Rössler systems. Phys. Lett. A **290**, 139 (2001)

# Erratum to: Predictability of Chaotic Dynamics: A Finite-time Lyapunov Exponents Approach

Juan C. Vallejo and Miguel A.F. Sanjuan

**Erratum to:**

**Juan C. Vallejo et al.,** Predictability of Chaotic Dynamics: A Finite-time Lyapunov Exponents Approach,
**Springer Series in Synergetics,**
**DOI 10.1007/978-3-319-51893-0**

Inadvertently the sources of the figures were not included originally. It has been included now for the below listed figures.

Chapter 1: For Fig. 1.2 and 1.3
Chapter 2: For Fig. 2.3 to 2.11
Chapter 3: For Fig. 3.1 to 3.13
Chapter 4: For Fig. 4.1 to 4.19

In the above listed figures, references have been cited in the figure legends. The corresponding references have been added at the end of the respective chapters.

---

The updated original online version for this book can be found at
DOI 10.1007/978-3-319-51893-0

---

Juan C. Vallejo
Department of Physics
Universidad Rey Juan Carlos
Móstoles, Madrid, Spain

Miguel A.F. Sanjuan
Department of Physics
Universidad Rey Juan Carlos
Móstoles, Madrid, Spain
ISSN

© Springer International Publishing AG 2017                         E1
J.C. Vallejo, M.A.F. Sanjuan, *Predictability of Chaotic Dynamics*,
Springer Series in Synergetics, DOI 10.1007/978-3-319-51893-0_5

# Appendix A
# Numerical Calculation of Lyapunov Exponents

## A.1 The Variational Equation

There is a large variety of numerical schemes for calculating the Lyapunov exponents. Each method possesses certain advantages and disadvantages, and a detailed comparison can be found in [13]. The most common methods can be divided into two main classes or families. Those based on the direct calculation of the distance between trajectories and those based on solving the variational equation.

The first family of methods, named as of *differences* or *direct*, derives from the definition of exponents as indicators of the divergence between points initially close enough. These schemes start by selecting two very close points, separated by a distance $d_0$. The system is iterated and the new separation is computed. The logarithm of the new and old separation is calculated and the scheme is repeated, averaging the obtained results.

For flows, the distance between trajectories can be easily computed as the phase space distance, but it would be worth a discussion on which norm could be used for the computation of the distance. For maps, the distance can also be calculated after a given number of iterations.

These methods imply a renormalisation process. That is, the distance should be set again to be the initial one after every certain number of steps. The reason is that when dealing with orbits within an attractor, the orbits do not diverge at a certain time scale, and may even begin to converge.

For flows, the renormalisation is done every $t$ time units. This means the existence of a scale factor that when multiplied by the distance $d(t)$ will return the original distance $d_0$. The asymptotic Lyapunov exponent can be obtained from averaging the logarithm of the new and old separation after applying the scale factor. For the calculation of all exponents and not only the largest one, the normalisation process is somehow more complex [10]. Basically, the computation of the second exponent can be done by considering the evolution of a two-dimensional surface evolving

© Springer International Publishing AG 2017
J.C. Vallejo, M.A.F. Sanjuan, *Predictability of Chaotic Dynamics*,
Springer Series in Synergetics, DOI 10.1007/978-3-319-51893-0

with $e^{(\lambda_1+\lambda_2)t}$. So, $\lambda_2$ can be computed once we have $\lambda_1$. The remaining exponents can be derived following the same idea.

One major disadvantage of the differences method is the selection of a practical initial distance $d_0$, and also a proper selection of the renormalisation period $t$. Since the strict definition requires infinitesimal deviations, to take finite initial deviations may lead to wrong results. It may happen that starting close to a limit cycle of exponent zero, the solution could be deviated towards initial conditions where the flow may converge. Conversely, when the points are as farther from the attractor as possible, there could be saturation effects. As a consequence, they cannot move away any farther, and the distance may keep roughly constant.

As a consequence of the above, nowadays it is more common to use the variational methods. The finite-time Lyapunov exponents are computed by solving the variational equation, that reflects the growth rate of the orthogonal semiaxes (equivalent to the initial deviation vectors) of one ellipse centred at the initial position as the system evolves [2]. The variational equation is essential when analysing the stability of orbits, evolution of phase space volumes under the dynamics, stable and unstable manifolds, and, obviously, Lyapunov exponents.

We define here $\boldsymbol{\Phi}(x, t)$ the solution of the flow equation. The time evolution of a phase-space point subject to a given flow dynamics is given by flow equations $\dot{x} = \boldsymbol{\Phi}(x, t)$.

Without loss of generality, we can detail an example based on the equations for a three-dimensional continuous flow:

$$\begin{cases} \dot{x} = f_1(x, y, z) \\ \dot{y} = f_2(x, y, z) \\ \dot{z} = f_3(x, y, z) \end{cases} \tag{A.1}$$

with initial condition $x_0 = (x_0, y_0, z_0)$.

Once the initial condition $x_0$ is fixed, we can integrate the flow during a given time $t$ and the initial point will follow certain trajectory in the phase space, ending in a final point $x$.

Imagine we add a small perturbation to $x_0$ in, say the $x$-direction. Evidently, the resulting initial perturbed condition vector will evolve towards a different point $x'$. The same can be said if we perturb the initial condition in other directions ($y$ and $z$, respectively).

The slopes of the flow in each direction provide a mean to know how the perturbation will evolve. It may be kept constant, enlarged, shrinked or even both, as it happens when the perturbation points out diagonally from a saddle point.

The matrix describing these slopes is the Jacobian matrix of the flow $\boldsymbol{\Phi}$, $\boldsymbol{J}$, that describes the evolution of deformations after a finite time $t$. So, $\boldsymbol{J} = D_v \boldsymbol{\Phi}$, contains the differential slopes in every possible direction,

$$\boldsymbol{J} = D_v \boldsymbol{\Phi} = \begin{bmatrix} \frac{\partial f_1}{\partial x} & \frac{\partial f_1}{\partial y} & \frac{\partial f_1}{\partial z} \\ \frac{\partial f_2}{\partial x} & \frac{\partial f_2}{\partial y} & \frac{\partial f_2}{\partial z} \\ \frac{\partial f_3}{\partial x} & \frac{\partial f_3}{\partial y} & \frac{\partial f_3}{\partial z} \end{bmatrix}, \tag{A.2}$$

We can use the Jacobian $\boldsymbol{J}$ for analysing how the perturbations, or variations, evolve under the flow dynamics.

The variations in each direction $[\delta_x]$, $[\delta_y]$ and $[\delta_z]$ are defined as vectors that will track the perturbation along each direction, as follows:

$$[\delta_x] = \begin{bmatrix} \delta_{xx} \\ \delta_{xy} \\ \delta_{xz} \end{bmatrix}, \tag{A.3}$$

$$[\delta_y] = \begin{bmatrix} \delta_{yx} \\ \delta_{yy} \\ \delta_{yz} \end{bmatrix}, \tag{A.4}$$

$$[\delta_z] = \begin{bmatrix} \delta_{zx} \\ \delta_{zy} \\ \delta_{zz} \end{bmatrix}. \tag{A.5}$$

The variation $[\delta]$ is then defined as the nine-component tensor,

$$[\delta] = \begin{bmatrix} \delta_{xx} & \delta_{yx} & \delta_{zx} \\ \delta_{xy} & \delta_{yy} & \delta_{zy} \\ \delta_{xz} & \delta_{yz} & \delta_{zz} \end{bmatrix}, \tag{A.6}$$

and the variational equation is then

$$[\dot{\delta}] = \boldsymbol{J}[\delta] = D_v \boldsymbol{\Phi}[\delta]. \tag{A.7}$$

Equivalently,

$$[\dot{\delta}] = \begin{bmatrix} \dot{\delta}_{xx} & \dot{\delta}_{yx} & \dot{\delta}_{zx} \\ \dot{\delta}_{xy} & \dot{\delta}_{yy} & \dot{\delta}_{zy} \\ \dot{\delta}_{xz} & \dot{\delta}_{yz} & \dot{\delta}_{zz} \end{bmatrix} = \begin{bmatrix} \frac{\partial f_x}{\partial x} & \frac{\partial f_x}{\partial y} & \frac{\partial f_x}{\partial z} \\ \frac{\partial f_y}{\partial x} & \frac{\partial f_y}{\partial y} & \frac{\partial f_y}{\partial z} \\ \frac{\partial f_z}{\partial x} & \frac{\partial f_z}{\partial y} & \frac{\partial f_z}{\partial z} \end{bmatrix} \begin{bmatrix} \delta_{xx} & \delta_{yx} & \delta_{zx} \\ \delta_{xy} & \delta_{yy} & \delta_{zy} \\ \delta_{xz} & \delta_{yz} & \delta_{zz} \end{bmatrix}. \tag{A.8}$$

So, in practical terms, the variational equation following the same pattern that the original equation flow is encoded by writing

$$[\dot{\delta}_x] = \begin{bmatrix} \dot{\delta}_{xx} \\ \dot{\delta}_{xy} \\ \dot{\delta}_{xz} \end{bmatrix} = \begin{bmatrix} \frac{\partial f_x}{\partial x} & \frac{\partial f_x}{\partial y} & \frac{\partial f_x}{\partial z} \\ \frac{\partial f_y}{\partial x} & \frac{\partial f_y}{\partial y} & \frac{\partial f_y}{\partial z} \\ \frac{\partial f_z}{\partial x} & \frac{\partial f_z}{\partial y} & \frac{\partial f_z}{\partial z} \end{bmatrix} \begin{bmatrix} \dot{\delta}_{xx} \\ \dot{\delta}_{xy} \\ \dot{\delta}_{xz} \end{bmatrix}, \tag{A.9}$$

for solving the evolution of the three-dimensional variation $\delta_x$. Following the same approach we have

$$[\dot{\delta}_y] = \begin{bmatrix} \dot{\delta}_{yx} \\ \dot{\delta}_{yy} \\ \dot{\delta}_{yz} \end{bmatrix} = \begin{bmatrix} \frac{\partial f_x}{\partial x} & \frac{\partial f_x}{\partial y} & \frac{\partial f_x}{\partial z} \\ \frac{\partial f_y}{\partial x} & \frac{\partial f_y}{\partial y} & \frac{\partial f_y}{\partial z} \\ \frac{\partial f_z}{\partial x} & \frac{\partial f_z}{\partial y} & \frac{\partial f_z}{\partial z} \end{bmatrix} \begin{bmatrix} \dot{\delta}_{yx} \\ \dot{\delta}_{yy} \\ \dot{\delta}_{yz} \end{bmatrix}, \tag{A.10}$$

for solving the evolution of the three-dimensional variation $\delta_y$, and similarly,

$$[\dot{\delta}_z] = \begin{bmatrix} \dot{\delta}_{zx} \\ \dot{\delta}_{zy} \\ \dot{\delta}_{zz} \end{bmatrix} = \begin{bmatrix} \frac{\partial f_x}{\partial x} & \frac{\partial f_x}{\partial y} & \frac{\partial f_x}{\partial z} \\ \frac{\partial f_y}{\partial x} & \frac{\partial f_y}{\partial y} & \frac{\partial f_y}{\partial z} \\ \frac{\partial f_z}{\partial x} & \frac{\partial f_z}{\partial y} & \frac{\partial f_z}{\partial z} \end{bmatrix} \begin{bmatrix} \dot{\delta}_{zx} \\ \dot{\delta}_{zy} \\ \dot{\delta}_{zz} \end{bmatrix}, \tag{A.11}$$

for solving the evolution of the three-dimensional variation $\delta_z$.

Therefore, we need to solve the variational equation and the system equation at the same time, working with the so-called augmented vector. For an $n$-dimensional system, this means to deal with a vector of $(n + n^2)$ variables. The first $n$ variables correspond to $n$ components of the "physical" $n$-dimensional vector, $x$, and the following $n^2$ variables are required for solving the evolution of the $n$-variations, each one being an $n$-dimensional vector.

## A.2   Selection of Initial Perturbations

By solving at the same time the flow equation and the fundamental equation of the flow, that is, the distortion tensor evolution, we can follow the evolution of the vectors, or axes, along the trajectory, and in turn, their growth rate. This method is described in [4] and [1].

The key point is that solving the variational equation implies solving the flow equations. So, for the augmented vector it must be selected a suitable initial condition, visualised as the initial axes lengths and directions of an initial deviation vector.

This is a key issue when dealing with finite integrations, because depending on this selection, the evolution of the distortion vectors will be different.

Obviously, there are several choices for the initial orientation of the ellipse axes. Due to the dependence on the finite integration time interval used in Eq. (2.11), every orientation will lead to different exponents [18].

A first simple choice may be the identity matrix, but this option does not seem to reflect any property of the flow. To allow the flow to point this initial selection to the most growing directions, and to obtain proper averaged indexes, we should integrate during long-times.

Another suitable option is to have the axes pointing to the local expanding/contracting directions, given by the eigenvectors of the Jacobian matrix. At local time scales the eigenvalues will provide insight on the stability of the point. Furthermore, these finite-time exponents can trace the stable and unstable manifolds, the latter with a time backwards integration [5, 7]. Note that in turn, the angle of both manifolds provides the nonhyperbolic nature of the system.

Another way of doing it could be to point the axes towards the direction which may have grown the most under the linearized dynamics, or to point them to the globally fastest growing direction.

Another possibility is to select the initial deviation vectors, by using the singular vectors linked to the Singular Value Decomposition (SVD) of the Jacobian matrix. The SVD takes into account that every matrix can be written in form of the product of three matrices. The first one is one formed by the left singular vectors (gene coefficient vectors, visualised as a "hanger" matrix). The second one is a diagonal matrix containing the so-called singular-values (mode amplitudes, visualised as a "stretcher" matrix). The third one is formed by the right singular vectors (expression level vectors, visualised as an "aligner" matrix).

The options listed above provide interesting insights on the behaviour of the dynamics of the flow at local scales when integrated during small finite time intervals. They are selected for pointing to directions that are already known "a priori" to express these local properties. As the integrations are larger, the local properties are washed out and the deviation vectors will end following the averaged global properties of the flow.

The axes will tend towards the fastest growing direction, may be at exponential rate, making their computation difficult to tackle. Because of that, the most commonly used methods use a Gram–Schmidt orthonormalisation process [3, 16]. By annotating the vector magnitudes before the normalization, we can calculate all Lyapunov exponents as defined in Sect. 2.1.

However, sometimes we are not merely interested in the final asymptotic values, but conversely, in looking for properties of the flow before the asymptotic values are returned. This is achieved by using the finite-time Lyapunov exponents, as described in Sects. 2.3 and 2.4.

It is worth to note that by selecting "a priori" directions, we may be already favouring the evolution towards the most growing directions. Therefore, some information about the time scales taken by the system for such an evolution may be lost.

As a consequence, there is also another interesting choice for the initial axes of the ellipse, that is to arbitrarily set them coincident with a random set of orthogonal vectors. This is the option selected along this book, used in, for instance, [14, 15].

With this selection, as the flow evolves, the axes get orientated from the arbitrary chosen direction as per the flow dynamics. Because this initial orientation will not favour any initial privileged direction, the growth rates of the axes of the ellipse will depend naturally on the flow time scales, and will begin to point to globally growing directions once a necessary finite time interval elapses.

Obviously, as the finite time interval grows above these time scales, the asymptotic regime will begin to appear. This algorithm returns the nearly asymptotic Lyapunov values ordered from the largest to the smallest when large enough time intervals are used. Conversely, there will be just a simple relationship among the exponents when using very local time scales [15]. But when using intermediate interval sizes, the returned values will characterise a given orbit.

## A.3   Other Methods

The two methods described earlier are the most commonly used. But it is worthy to mention the existence of additional algorithms and methods. We refer the interested reader to follow, for instance, [13] and [9].

There are the so-called QR methods, based on the factorisation of the matrix resulting from the QR decomposition of the Jacobian, writing it as the product of an orthogonal matrix Q and an upper triangular matrix R. These methods seem to be more adequate for the computation of LCEs, but not for the computation of FTLEs, because they introduce certain errors that are only cleared as the integration time increases. There are also methods based on the SVD decomposition mentioned in the previous section. In general, both methods could not be so convenient in nonhyperbolic systems. In these systems, the FTLEs can accumulate around zero and the decomposition to have an almost degenerate spectrum. Certainly, there are corrections to the QR method that can be applied to degenerated cases. And indeed, there are modifications for proper handling of Hamiltonian systems, incorporating their symplectic nature [11]. Nevertheless, some of these modifications are only effective in systems with one or two degrees of freedom, and they are not very suitable for systems with a higher number of degrees of freedom. In any case, every method has certain advantages and disadvantages, and some additional description can be found in [13] and [12].

A final remark could be given regarding the computation of Lyapunov exponents in time series. In these cases some specific methods are required [6, 17]. If we do not have an appropriate knowledge of the fundamental equations of the system, having only experimental data available, these methods can be classified into two families: direct methods or tangent space methods.

The direct methods are based on searching for time series in the neighbourhood of the initial point, and rely on computing the necessary comparisons. The tangent space methods perform the computation by predicting the Jacobian using the available time series. See [8] and the references therein.

## A.4 Practical Implementation for Building the Finite-Time Distributions

If we make a partition of the whole integration time along one orbit into a series of time intervals of size $\Delta t$, then it is possible to compute the finite-time Lyapunov exponents $\chi(\Delta t)$ for each interval.

The distribution is built by integrating the augmented vector under the flow dynamics up to a selected $\Delta t$ interval. We fix the initial point of the orbit, as desired, and as initial perturbation an arbitrary set of orthonormal vectors, as described in the previous Sect. A.2. We will keep it for later use.

A small note could be raised here related to this orthonormalisation process. A widely used algorithm is the Gram–Schmidt process. This has been considered as inherently numerically unstable, that is, very sensitive to round-off errors. We can alleviate this by avoiding divisions by small numbers and column pivoting appropriately. We should aim to more numerically stable processes by using some of the available modified versions of the Gram–Schmidt algorithms or by using Householder transformations.

When integrating, at each integration step, we propagate the variations, calculating how the log of their norms evolve following Eq. (2.11), and Sect. 2.4.

When the finite time value is reached, we save the value of the calculated finite-time exponent $\chi(\Delta t)$, and start the cycle again, resetting the augmented vector. The new initial condition is the current point of the trajectory, when we have stopped. The new initial perturbation will be the one we selected previously. This is done to assure that we are comparing how the same perturbation evolves along the trajectory points.

We repeat the above process until the total integration time is reached. This integration time could be as long as needed, embracing any transient period, and going farther, or stopping the integration once the transient has ended.

This distribution of finite-time Lyapunov exponents can be normalized dividing it by the total number of intervals, thus obtaining a probability density function $P(\chi)$, that gives the probability of getting a given value $\chi$ between $[\chi, \chi + d\chi]$.

There is a huge amount of available languages and mathematical packages for solving the dynamical flow equation, the variational equation, and computing the finite-time or asymptotic infinite Lyapunov exponents. Some of them have already implemented many of the required algorithms, and the final choice for using a given language or package, commercial or free is up to the user.

# References

1. Alligood, K.T., Sauer, T.D., Yorke, J.A.: Chaos. An Introduction to Dynamical Systems, p. 383. Springer, New York (1996)
2. Alligood, K.T., Sauer, T.D., Yorke, J.A.: Chaos. An Introduction to Dynamical Systems, p. 195. Springer, New York (1996)
3. Bay, J.S.: Fundamentals of Linear State Space Systems. McGraw-Hill, Boston (1999)
4. Benettin, G., Galgani, L., Giorgilli, A., Strelcyn, J.M.: Lyapunov characteristic exponents for smooth dynamical systems and for Hamiltonian systems; a method for computing all of them. Meccanica **9**, 20 (1980)
5. Doerner, R., Hubinger, B., Martiensen, W., Grossmann, S., Thomae, S.: Stable manifolds and predictability of dynamical systems. Chaos Solitons Fractals **10**, 1759 (1999)
6. Eckmann, J.P., Oliffson, S., Ruelle, D., Ciliberto, S.: Lyapunov exponents from time series. Phys. Rev. **34**, 4971 (1986)
7. Joseph, B., Legras, B.: On the relation between kinematic boundaries, stirring and barriers. J. Atmos. Sci. **59**, 1198 (2002)
8. McCue, L.S., Troesch, A.W.: Use of Lyapunov exponents to predict chaotic vessel motions. In: Contemporary Ideas on Ship Stability and Capsizing in Waves. Fluid Mechanics and Its Applications, 1, vol. 97, Part 5, pp. 415–432. Springer, Dordrecht (2011)
9. Okushima, T.: New method for computing finite-time Lyapunov exponents. Phys. Rev. Lett. **91**, 25 (2003)
10. Parker, T.S., Chua, L.O.: Practical Numerical Algorithms for Chaotic Systems. Springer, New York (1989)
11. Partovi, H.: Reduced tangent dynamics and Lyapunov spectrum for Hamiltonian system. Phys. Rev. Lett. **82**, 3424 (1999)
12. Ramasubramanian, K., Sriram, M.S.: Lyapunov spectra of Hamiltonian systems using reduced tangent dynamics. Phys. Rev. E **62**, 4850 (2000)
13. Ramasubramanian, K., Sriram, M.S.: A comparative study of computation of Lyapunov spectra with different algorithms. Physica D **139**, 72 (2000)
14. Vallejo, J.C., Sanjuan, M.A.F.: The forecast of predictability for computed orbits in galactic models. Mon. Not. R. Astron. Soc. **447**, 3797 (2015)
15. Vallejo, J.C., Viana, R., Sanjuan, M.A.F.: Local predictibility and non hyperbolicity through finite Lyapunov exponents distributions in two-degrees-of-freedom Hamiltonian systems. Phys. Rev. E **78**, 066204 (2008)
16. Wolf, A.: Quantifying chaos with Lyapunov exponents. In: Chaos, Chap. 13. Princeton University Press, Princeton (1986)
17. Wolf, A., Swift, J., Swinney, H., Vastano, J.: Determining Lyapunov exponents from a time series. Physica D **16**, 285 (1985)
18. Ziehmann, C., Smith, L.A., Kurths, J.: Localized Lyapunov exponents and the prediction of predictability. Phys. Lett. A **271**, 237 (2000)

Printed in the United States
By Bookmasters